# Superheated Steam Drying

Superheated steam drying (SSD) has long been recognized for several major advantages it offers over other convective dryers, including high energy efficiency by utilization of energy in the exhaust steam, higher product quality due to the absence of oxygen, and avoidance of fire and explosion hazards.

Offering a global critical overview of the current state of the art, *Superheated Steam Drying: Technology for Improved Sustainability and Quality* assesses future needs and opportunities for industry adoption and further innovation in SSD. It covers SSD technologies for various industrial sectors and mathematical modeling approaches to help with design and scale-up. The effects of SSD on drying kinetics as well as product quality are also discussed with examples.

This book serves as a useful reference for technicians, graduate students, and researchers in the field of drying technology. It can also be used in courses on industrial drying, processing and drying of food, advanced drying technology, and superheated steam drying.

**Advances in Drying Science and Technology**
*Series Editor:*
Arun S. Mujumdar,
*McGill University, Quebec, Canada*

Heat and Mass Transfer in Drying of Porous Media
*Peng Xu, Agus P. Sasmito, and Arun S. Mujumdar*

Freeze Drying of Pharmaceutical Products
*Davide Fissore, Roberto Pisano, and Antonello Barresi*

Frontiers in Spray Drying
*Nan Fu, Jie Xiao, Meng Wai Woo, and Xiao Dong Chen*

Drying in the Dairy Industry
*Cécile Le Floch-Fouere, Pierre Schuck, Gaëlle Tanguy, Luca Lanotte, and Romain Jeantet*

Spray Drying Encapsulation of Bioactive Materials
*Seid Mahdi Jafari and Ali Rashidinejad*

Flame Spray Drying: Equipment, Mechanism, and Perspectives
*Mariia Sobulska and Ireneusz Zbicinski*

Advanced Micro-Level Experimental Techniques for Food Drying and Processing Applications
*Azharul Karim, Sabrina Fawzia, and Mohammad Mahbubur Rahman*

Mass Transfer Driven Evaporation of Capillary Porous Media
*Rui Wu, Marc Prat*

Particulate Drying: Techniques and Industry Applications
*Jangam Vinayak, Chung-Lim Law, and Shivanand Shirkole*

Drying and Valorisation of Food Processing Waste
*Chien Hwa Chong, Rafeah Wahi, Chee Ming Choo, Shee Jia Chew, and Mackingsley Kushan Dassanayake*

Drying of Herbs, Spices, and Medicinal Plants
*Ching Lik Hii and Shivanand Shirkole*

Drying of Aromatic Plant Material for Natural Perfumes
*Viplav Hari Pise, Ramakant Harlalka, and Bhaskar Narayan Thorat*

Dried Fruit Products
*Felipe Richter Reis and Shivanand S. Shirkole*

Superheated Steam Drying: Technology for Improved Sustainability and Quality
*Mukund Haribhau Bade, Sachin Vinayak Jangam, and Arun S. Mujumdar*

**For more information about this series, please visit:** www.routledge.com/Advances-in-Drying-Science-and-Technology/book-series/CRCADVSCITEC

# Superheated Steam Drying

## Technology for Improved Sustainability and Quality

Mukund Haribhau Bade,
Sachin Vinayak Jangam, and
Arun S. Mujumdar

CRC Press is an imprint of the
Taylor & Francis Group, an informa business

First edition published 2025
by CRC Press
2385 NW Executive Center Drive, Suite 320, Boca Raton FL 33431

and by CRC Press
4 Park Square, Milton Park, Abingdon, Oxon, OX14 4RN

*CRC Press is an imprint of Taylor & Francis Group, LLC*



ISBN: 978-1-032-23027-6 (hbk)
ISBN: 978-1-032-23031-3 (pbk)
ISBN: 978-1-003-27529-9 (ebk)

DOI: 10.1201/9781003275299

Typeset in Times
by codeMantra

# Contents

# Figures

# Tables

# Abbreviations, Nomenclature, and Notations

**Abbreviations and Full Forms**

| | |
|---|---|
| **BSG** | Brewer's spent grain |
| **CDRR** | Constant drying rate region |
| **FB** | Fluidized bed |
| **FBD** | Fluid bed dryer |
| **FDRR** | Falling Drying Rate Region |
| **HA** | Hot air |
| **HAD** | Hot air drying/dryer |
| **LP** | Low pressure |
| **RF** | Radio frequency |
| **SEC** | Specific energy consumption |
| **SS** | Superheated steam |
| **SSD** | Superheated steam drying/dryer |
| **USSR** | Union of Soviet Socialist Republics |

**Nomenclatures**

| | |
|---|---|
| $D_0$ | Arrhenius factor |
| $D_{eff}$ | Effective diffusion co-efficient ($m^2/s$) |
| $E_a$ | Activation energy (kJ/mol) |
| $L$ | Maximum thickness (m) |
| $M$ | Moisture content (d.b.) (kg/kg of dried products) |
| $R$ | Universal gas constant (8.314 J/mol K) |
| $t$ | Drying time (s) |
| $T$ | Temperature (K/°C) |
| $x$ | Horizontal or $x$-axis for thickness (m) |

**Subscripts/superscripts**

| | |
|---|---|
| $e$ | Equilibrium |
| $i$ | Initial |

# Preface

Although the idea of using superheated steam as a convective drying medium in place of hot air or combustion gases to dry various materials was published in a German book almost 125 years ago, its first application in industry took place some four decades later, mainly to avoid fires during drying. It took some three more decades for new applications for flash-drying pulp to emerge. Over the past five decades, interest and research and development activities in this area increased significantly but also have not yet been fully exploited despite key advantages such as its low net energy consumption if the energy from dryer exhaust is recovered, less emissions from the dryers and reduced carbon emission (least water and carbon footprint), better-quality product because of no oxidative damage, avoidance of fire and explosion, and so on. On the adverse side, the drying system is more complex and capital intensive due to the need to recover energy, requirement of appropriate feed supply and dried product extraction system in continuous operation, and better sealing of the drying chamber with proper way out for intermittent operation.

This book is designed to address various issues related to superheated steam drying (SSD) technologies. We have critically reviewed the current state-of-the-art literature and identified important advantages and limitations, various drying equipment and drying products, energy aspects of the SSD, etc. This book includes a discussion of the basic principles, energy and product quality aspects, effect of process parameters such as pressure and temperature, and so on. A number of real-world case studies are included as well along with ideas for further enhancements to this technology, where relevant comparison is made with conventional drying with air and other technologies. Detailed discussion is included of various approaches to mathematical modeling of SSD. A quick perusal of the contents will allow the reader to recognize the wide potential of SSD for products sensitive to oxidative damage with additional advantages of product quality due to unique way of drying kinetics.

Indeed, its potential in drying very large-volume commodities such as coal, biomass, various sludge, paper pulp, wood, and so on is already established. Many more products need to be added to this growing list. Further, the senior author Prof. Mujumdar has significant contribution to the development of both science and technology involved in SSD since 1978 with a few novel ideas, which are definitely useful to make content rich and maintain quality. We have attempted to make this book a valuable sourcebook for researchers as well as industrial users of dryers and vendors of such equipment.

We hope that this concise monograph will encourage new research in this challenging area—often touted as drying technology of the future. As the cost of energy rises and legislative pressure on all energy-intensive processes using fossil fuels intensifies on a global scale in addition to better-quality products we believe more attention will be paid to SSD in the coming decades.

# Acknowledgments

The book project entitled *Superheated Steam Drying: Technology for Improved Sustainability and Quality* is the first and only monograph in English, focused on superheated steam as the drying medium. During this project, many friends, students, researchers, and faculty members helped us to bring it to fruition.

We thank the CRC reviewers who provided valuable suggestions to enhance our proposal. Many eminent researchers wholeheartedly shared case studies which added value to the content of this monograph. Prof. Sankar Bhattacharya of Monash University willingly provided information for a case study on low-rank coal drying. Dr. Arne Jensen, Founder, Partner EnerDry A/S, Denmark, consented to include the case study on a high-pressure superheated steam fluidized bed dryer for beet pulp. We are grateful to Dr. Ing Antoine Dalibard of Fraunhofer Institute for Interfacial Engineering and Biotechnology IGB, Germany, for his permission to use some artwork and for helping us with information on dryers manufactured by Fraunhofer IGB, Germany.

Furthermore, we extend our special thanks to Dr. Aparupa Pani, Dr. Savitha Srinivasan, and Dr. Viplav Pise of ICT, Mumbai, India, who reviewed several final drafts and provided valuable feedback. Our heartfelt thanks also go to the following publishers for granting permission to use copyrighted material: Taylor and Francis, Elsevier, and Springer Nature. We appreciate the production staff of CRC Press for their continuous guidance and assistance throughout the progress of this project.

Finally, writing a book requires a very significant amount of time and effort that is beyond that required for academic duties; this comes from time that we could not devote to our families. We are therefore grateful to our families for their support, which made it possible for us to complete this project, albeit with significant extension beyond our original optimistic projection of the completion date.

# About the Authors

**Mukund Haribhau Bade** holds a PhD in energy modeling and process integration from the Department of Energy Science and Engineering, IIT Bombay, Mumbai, India. He has over 24 years of teaching and research experience in various technical institutes. He is currently working as an associate professor at the Department of Mechanical Engineering, Sardar Vallabhbhai National Institute of Technology, Surat, Gujarat, India. His main research interests are energy management, energy analysis and modeling, pinch analysis, drying technology including superheated steam drying, and pump as turbine for micro-hydro applications. He has published more than 50 articles in reputed international journals and conferences. He works as a reviewer for many reputed journals, like *Energy Conversion and Management*, *Drying Technology*, *Clean Technologies and Environmental Policy*, and *ISH Journal of Hydraulic Engineering*. In addition, he is also working on funded projects related to energy analysis and modeling for performance enhancements including textile dryers. He has mentored over four PhD students and guided projects of many master and undergraduate students.

**Sachin Vinayak Jangam** is a senior lecturer in the Department of Chemical and Bimolecular Engineering at the National University of Singapore (NUS). He completed his PhD in Chemical Engineering at the Institute of Chemical Technology, Mumbai, India. He has worked on mathematical modeling and experimental analysis of industrial drying of various products as a major part of his PhD thesis. He then worked as a research fellow at the Minerals, Metals and Materials Technology Center at NUS, developing cost-effective drying techniques for minerals. He has published several research articles, review papers, and book chapters on drying and related fields. He is a co-author of a book on foundational concepts of chemical engineering and edited several free e-books on drying. His current work focuses on drying, energy minimization, and pedagogy in chemical engineering education. He has been part of the editorial team of the archival journal *Drying Technology* (Taylor & Francis) since 2015.

**Arun S. Mujumdar** holds a PhD in Chemical Engineering from McGill University, Canada, and Doctor Honoris Causa from Lodz Technical University, Poland, University of Lyon, France, and Western University, Canada. He was a professor of Mechanical Engineering at the National University of Singapore (NUS), following a long tenure as a professor of Chemical Engineering at McGill University, Canada. He has mentored over 100 PhD students at several universities and published over 600 peer-reviewed papers which have attracted over 55,000 citations. He is the author of three books and 70 edited books on heat and mass transfer, drying, and mathematical modeling in transport processes. He holds honorary professorships in several universities. His research areas are transport phenomena, thermal management of fuel cells and battery stacks, energy systems, dewatering and drying, and innovative drying processes in diverse industries, including foods, paper, products of biological origin, minerals, pharmaceuticals, and sludge.

# 1 Introduction

## 1.1 OVERVIEW

Drying is one of the oldest technologies used by human beings since immemorial mainly for food and other things to be preserved. It can be defined as the removal of a fixed quantity of water as vapor from the feedstock available as liquid, semi-solid, paste, suspension, or wet solid by various means such as applying thermal energy, dielectric, freeze drying, radio frequency (RF), or microwave, etc. under controlled conditions [1]. These are required for any or more reasons, viz., to extend the shelf life of products, reduce weight for economical transportation or disposal, require intermediary product in value addition chain, or removal and disposal of pollutants by environmentally sustainable approach, etc. The details of each drying technology with energy sources, physics of drying, and mode of water removal are explained in various drying books, notably Mujumdar [1], Law et al. [2], and Rahman [3]. The most common technology for drying is thermal energy transfer by convection heat transfer through hot air (HA), combustion (flue) gases, or superheated steam (SS). In hot air drying (HAD), atmospheric air is heated either directly (mixture of air and combustible) or indirectly in a closed chamber and pass it on to the feedstock to be dried in a drying chamber, may be in continuous or batch operation. The dried product is taken out appropriately and exhaust moist air may be recirculated partially as heat recovery after cleaning it. These hot air (HA) based dryers have well-developed technology available for a huge number of products and varieties due to their unique advantages of easy availability and adaptability with diverse products; however, it may not be ideal drying media for all the feedstocks. Hot air as drying media has very poor heat transfer properties, making the size of the dryer bulky and high operating cost with pollution issues due to huge energy consumption and emission of air with other pollutants. Further, being high-oxygen content, drying with combustible or oxidizing products is challenging. On the other hand, superheated steam drying (SSD) has a long history of applications as drying media from the beginning of 20th century [4], but its commercial applications are recent and that too for a few products like wood chips, hog fuel, beet pulp, various food products, etc. [5–7]. In this, SS (steam above the saturation temperature at drying chamber pressure) is used as a heat source to evaporate the moisture present in wet feedstock by convection heat transfer mode using only sensible heat (superheat horn) of steam and to drive evaporated water generated due to drying making its part. This technology is in a developing stage where many more areas still need to be explored. This chapter deals with historical development of SSD, the physical mechanism of moisture removal, advantages, and limitations with detailed applications.

DOI: 10.1201/9781003275299-1

## 1.2 HISTORICAL DEVELOPMENT

The concept of drying using SS was first proposed at the beginning of 20th century by Hausbrand [4] for below, above, and at atmospheric pressure drying process with the feasible layout of various components involved as well as the possibility using initial calculations of the requirement of SS. This book was initially published in German in 1898 [4], and afterward, there were English translations as the first edition in 1901, the second edition in 1912, and so on. Hausbrand [4] also emphasized various notable advantages and limitations of superheated steam drying, which are still relevant. However, the early reported commercial dryer based on SS was originated in 1920 by the Swiss engineer Karrer [8]. This was based on SS as a drying medium and an electrical heater as the heat source for the oven, where peat, vegetables, and fruits were dried. He had proposed three designs of early SSD called ovens at three different temperatures as low temperature (below 100°C) working at vacuum pressure, medium temperature (between 100°C and 200°C) at near atmospheric pressure, and high temperature (above 200°C) at high pressure (above atmospheric pressure). In this case, high-temperature ovens used to dry the foundry-molds and cores with SS were able to get an efficiency of 37%, which was almost double that of HA ovens used at that time. The next application tried is for peat with a maximum temperature 130°C where efficiency reached 82% due to the recirculation of SS. The third and last application was from biomaterials, mainly the drying of cabbages and grass reported in the same paper. The cabbage was dried using low-pressure superheated steam dryer (LPSSD) with a temperature maximum of up to 130°C and pressure 700mm of mercury column (Hg). It took a total 5h, and it is important to note that the color of the cabbage was retained as green. Similarly, grass was also dried in 5h as green hay. These all operations are performed for batch dryers, where a certain amount of particular feed is dried in the drying chamber. The purpose of the work was to show the potential of SSD and hence no internal details of SSD are documented.

Other early efforts for SSD applications are the dehydration of coal developed by Fleissner [9] in 1920 and it is rewarded by his name as the Fleissner process. Though it is not a true SSD process, as flashing of pressurized saturated steam results in the removal of moisture from coal, it is referred due to its close relevance. In this process, coal is lumped into a drying chamber and pressurized till 12–12.5 bar with saturated steam. The charge is thoroughly heated up to the required temperature using steam and then pressure is reduced sometimes may be below atmosphere such that water in the form of moisture available in the coal flashes to vapor. Walker et al. [10] commented in his book published in 1937 "This method of drying (using superheated vapor) seems capable of wide application…", which signifies the fact that regardless of known benefits, drying medium as superheated vapors was not extensively applied. The reasons may be the SSD was considered suitable for solids which are sustaining the operating temperature of the superheated vapor at atmospheric pressure (100°C), depending on the necessity of equilibrium moisture (degree of dryness), and it is hard to achieve low moisture after completion of drying operation [11]. For batch operations of the SSD, huge start-up and shut-down losses were predicted and asserted that to overcome it, costly equipment may be needed, limiting further applications of it. However, after the Second World War, researchers have focused on the applications of SSD to take advantage of energy economy and also proved that, compared to HAD, SSD has various advantages in addition to energy economy, recirculation, and recovery.

Eisenmann [12] proposed the theory of high-temperature lumber drying in SS and noticed that SSD is cheaper and quicker than HAD. However, excessive drying resulted in poor quality of lumber; also corrosion of the drying equipment discouraged the adoption of the process by the lumber industry. Other work by various researchers showed that, when compared between HAD and SSD for granular material as sand, the latter had higher drying rates and thermal efficiency, outweighing the higher capital costs due to savings in operational expenses [13]. Dungler [14] was granted a patent for impinging jet SSDs, which were intended to be used in the production of paper, fabric, and other fibrous sheet materials. However, the patent did not address heat recovery from steam recirculation or excess steam that was evaporated and released into the atmosphere; it is possible that this technology was never commercialized. Beane [15] proposed dehydration of alfalfa with SS, which is not only beneficial for lower energy consumption but also retains vitamins as carotene of it compared to HAD. The pilot trial of SSD such as tray and turbotype dryers for wet filter cake and granular materials [16] with significant advantages other than energy savings as well as comparison with HAD are described in detail by Lane and Stern [17].

The use of SS for wood drying began in the initial year of $20^{th}$ century; however, due to lack of detailed understanding and physics of SSD, commercial development was halted, which was again picked up after Second World War as discussed by Kauman [18]. Equilibrium moisture equations for various conditions of pressure below, at, and above the atmosphere were given for wood with the help of psychrometric charts. Overdrying of the wood adversely affects the quality, and hence precise control of drying in SS is necessary [18].

On the fundamental side, several attempts have been reported in the English literature on comparative studies of direct drying in steam versus air. Wenzel and White [13] provided quantitative measurements of air and steam drying of beds of granular solids like sand. Chu et al. [19] investigated the formation of steam from water by evaporation and humid air and reported that the drying rate is in decreasing order with the medium as SS, a mixture of HA and SS (humid air), and hot dry air, respectively; but for humid air with more than 50% SS, the drying was not beneficial in both constant and falling rate periods for medium temperatures. Malmquist and Noack [20] studied steam drying of hardwoods (e.g., beech and oak) and demonstrated faster drying in steam than HA. Hann [21] studied high-pressure steam drying of yellow poplar at pressures up to 5 bar and showed that darkening as well as honeycomb collapse occurs in this case.

Yoshida and Hyodo [22] performed exhaustive experimentation for SS, humid air, and HA dryers with food products and first time reported that high drying rates allow higher retention of nutrients. They were probably the first to define and verify systematically the "inversion temperature," the drying medium temperature below which the rate of evaporation falls as the moisture increases and above which the reverse is true for SS. The proposed relations were confirmed by the experiments performed using a wetted-wall column to maintain the strict convection heat transfer mode dominant for all dry air, wet water vapor, and SS and based on it inversion temperature reported was around 175°C.

There is no textbook devoted to SSD in English till date, but in 1967, Mikhailov [23] from Russia (erstwhile USSR) published a 200-page book entitled *Superheated Steam Drying* in Russian. From a cursory examination of the contents of this book, it appears that the steam drying concept had more significant followers in the USSR than in the West, although much of the work was related to laboratory studies, including the consequences of drying using steam on the structure and properties of the materials, an area of considerable current interest everywhere owing to its implications on product quality. The only industrial process discussed in this book published in 1967 is the drying of peat in steam dryers—a process that was adopted in the West only in the late 1970s. It appears that steam drying of coal has been practiced in East Germany since the early 1970s although access to relevant work is rather restricted.

After 1980, a number of researchers started working on the development of diverse applications of SSD at laboratory, pilot, and commercial stages. But the real pace of publications on laboratory, pilot, and industrial trials was picked up only after 1990. Table 1.1 summarizes the commercial development of the SSD till 2023.

## 1.3 WORKING PRINCIPLE OF SUPERHEATED STEAM DRYER

The thermal energy required for the drying of wet feedstock is supplied by creating the SS environment, which transfers it by mainly convection mode. Main components, such as a drying chamber, steam generator, superheater, blower for steam recirculation, heat recovery system, feed supply, dry product outlet, etc., are shown in Figure 1.1. Initially, the SS produced from the water in the steam generator and superheater is passed through the drying chamber, and feedstock is supplied appropriately into the drying chamber so that steam will not be leaked. Feedstock may be preheated to saturation temperature to avoid the initial condensation of the SS, but in some cases, initial condensation will be favorable [26]. The SS supplied the required thermal energy for the evaporation of the moisture both free and bound in the feedstock, reducing the temperature of SS up to just above saturation temperature. Initially, all free or surface moisture will be evaporated and then bound moisture will be transferred to vapor appropriately. The evaporated moisture immediately becomes part of the supply SS and will be exhausted from the drying chamber at a lower temperature, which can be partially recirculated back into drying chamber after cleaning and reheating to the supply temperature. The excess steam from the exhaust may be used in the plant elsewhere if possible, or its energy may be recovered appropriately. In this case, as the steam is recirculated back into the drying chamber, the energy supplied to SS is only sensitive heat and latent energy of evaporation of the steam will be possible to be recovered elsewhere as discussed above. Figure 1.2 depicts the various regions of the water-steam with pressure and temperature along with required heat duty. Because of these, the energy required will be significantly reduced for SSD compared to HAD. However, this is not true for all the temperatures ranges, for example, above inversion temperature (critical temperature), the SSD rate is higher than HAD and vice versa, which is explained in more details afterwards.

**TABLE 1.1**
**Commercially Developed Superheated Steam Dryers**

| S.N. | Products | Comments |
|---|---|---|
| 1 | Drying of bark, hog fuel, beet pulp, bagasse, wastewood, straw etc. | No emissions; no fire/explosion hazards. Short residence times, 5–60 s depending on product |
| 2 | Dietary fiber; black pepper, corn gluten, food products, etc. | Some cooking in steam environment. For vegetable products with >80% moisture. Product disinfected (even for dry product). Carbohydrates made more hydrophilic and rewettable; solubility of pectins improved, proteins made more digestible, etc. Heat consumption 180 kWh/ton water evaporation. Power consumption 40 kWh/ton water evaporation. Limitation: starch may gelatinize |
| 3 | Hog fuel drying (Flash dryer) | Drying of hog fuel before feeding on fuel to boilers improves boiler efficiency (e.g., steam production increased from 4.5 kg steam/kg dry solid to 6.1 kg steam/kg solid using a steam dryer). With conventional drying the improvement is 0%–5%. |
| 4 | Coal Drying (Fluidized bed) [24] | Installations in the USSR, Australia, South Africa, etc. Compact dryer. No fire/explosion hazards. Economics significantly better than air drying in rotary or fluidized beds. |
| 5 | Drying and deodorization of soy sauce cake (Fluidized bed dryer, agitated bed dryer) [25] | Uses indirect heating of dryer walls to reduce steam requirements. Steam blown through agitated bed of soy sauce cake. Drying and deodorization take place concurrently. Oil evaporating from cake burnt in fume incinerator reducing fuel consumption by 30%. Steam recycling found uneconomic for this application (2 T dry product/h). |
| 6 | Drying of textiles (Stenter dryer; Star Manufacturing Co., Bombay, India) | Essentially impingement dryer. Over 35% reductions in energy costs and better product quality. Operational since 1980. |
| 7 | Mineral fibers (replacement for asbestos as insulation) | New process developed in Sweden for the production of Rockwool. 99.5% final dryness; individual fibers are fluffier with better insulating properties. |
| 8 | Corn gluten fiber (Cattle feed) | Full-scale operation using Exergy Flash dryer since 1986. 30% reduction in steam consumption over conventional dryers. Light-colored, uncontaminated, fluffy product. Sticky, heat-sensitive material. |
| 9 | Drying of wet milled peat (using Vapor Recompression) Exergy dryer | New Swedish plant will dry 40% d.b. peat to 90% dryness. Two dryers—each 30 T/day evaporated water. No external consumer of steam in peat harvest areas so latent heat recovery will use two turbo compressors increasing pressure from 3 to 14 bar. Total power consumption including fans: 190 kWh/ton water evaporation. District heat of 3–6 MW supplied to nearby village. |

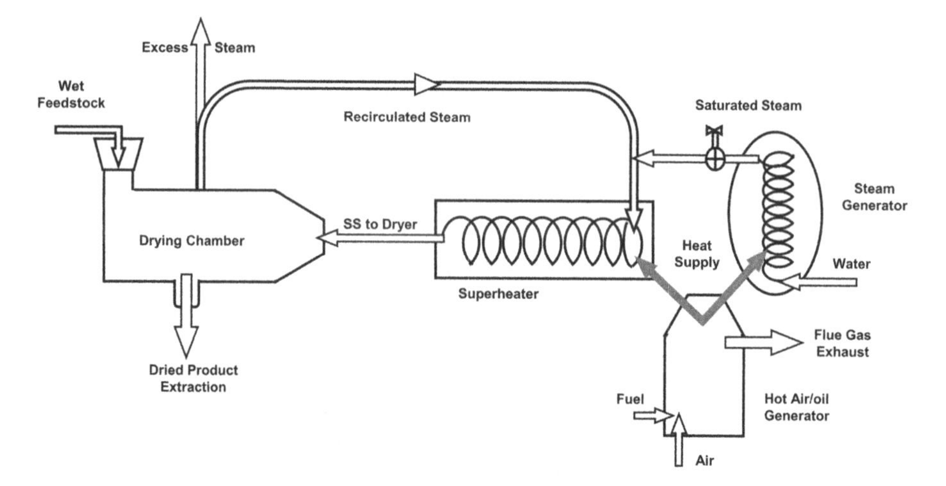


**FIGURE 1.1** Line diagram representation of superheated steam dryer with various components.

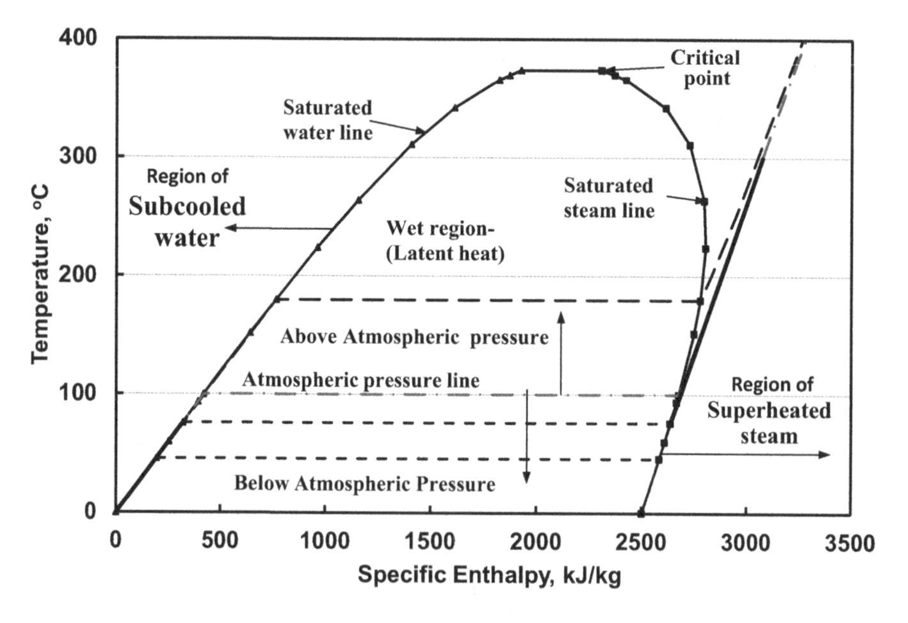


**FIGURE 1.2** Water-steam phase diagram with various regions: saturated water, vapor, and superheated steam with pressure.

## 1.4 ADVANTAGES AND DISADVANTAGES OF SSD OVER HAD

The primary advantages of SSD known from the beginning of 20th century and other more recent benefits specific to particular products [26–28] discovered later on are elaborated systematically as follows:

### 1.4.1 Primary Advantages

i. Energy efficient: Compared to HA, the reuse and heat recovery of SS albeit at small value of specific enthalpy are easy (due to higher saturation temperature and cleaning), effective (relatively at a higher temperature), and economical (conservation of the latent energy of evaporated water). Therefore, partial exhaust SS (at a lower temperature) can be easily cleaned and recirculated after sensibly reheating in a superheater. This conserves the energy otherwise lost in exhaust like HAD. Further, the excess SS approximately equal to the evaporated moisture can be appropriately used by thermocompression or mechanical compressor to uplift its specific enthalpy for recycle in the dryers or may have other applications in the plant either directly or indirectly recovering its energy. In such cases, latent heat content in SS will not be utilized for drying, resulting in specific energy consumption of 1000–1500 kJ/kg of water evaporation for SSD as opposed to 4000–6000 kJ/kg for HAD. Consequently, theoretically, a reduction of net energy consumption by around 75% is a net gain for the SSD, and in practice, this will be around 40%–50%, which is still significantly high [26]. If there is no use of excess steam in the plant and no recovery of the thermal energy, then economic advantages will not be availed.

ii. Due to better thermodynamic properties such as heat capacity, heat transfer coefficient, thermal conductivity, etc., for the same drying capacity, SSD is more compact than HAD. But the overall size of SSD is bigger than that of HAD due to more numbers of systems.

iii. Being SS environment, no oxygen (or a meager quantity) is present in the drying chamber, and it will result in no or unnoticeable oxidation or combustion reactions in SSD, which are detrimental in so many cases.

iv. For the above inversion temperature, SSD has a higher surface moisture removal rate (for constant drying rate region) than HAD mainly due to higher heat capacity and thermal conductivity of SS and, lower than the inversion temperature, it is vice versa. For the falling rate region, the product temperature is higher in SSD (more than 100°C at atmospheric pressure) and lower diffusion resistance for moisture evaporation (no air) results in greater drying rates for the entire temperature range.

v. Generally, recycling of the drying steam and recovery of heat energy from excess steam in SSD lead to a reduction of pollution due to emissions into the atmosphere, improving the green footprint.

vi. In the drying process, dried particles in small quantities are carried by the exhaust stream, which can be easily recovered in SSD as condensation in comparatively smaller condensers is possible. This is especially important for the products to be dried, containing poisonous or costly organic substances. Also being steam medium as non-oxygen or very little oxygen presence, there are no fire hazards or explosions possible.

vii. SSD allows pasteurization, deodorization, sterilization, etc., and adds certain desirable valuable features, such as porosity, hydrophilicity, etc., of food products.

### 1.4.2 Advantages for Food and Biomaterials

i. Due to an oxygen-free (negligible oxygen) environment, unwanted oxidation reactions such as enzymatic browning reactions, oxidative deterioration of vitamins, and lipid oxidation cannot take place in the SS drying medium. Further, in the case of biofuels, it restricts explosion risks, lowers the possibility of fires, and delivers products without any burn marks.
ii. Because of an oxygen-free environment, food preservation is easy [29] compared to conventional HAD because SS also helps for pasteurization, disinfection, and sterilization of them. The SSD can play a role in effectively inactivating microorganisms, which may contaminate food items to be dried. Generally, during SSD, feedstock temperature is above 100°C at atmospheric pressure (high temperature), which is good for inactivating most of the microorganisms. At the initial period of the drying, when the feedstock is supplied to a drying chamber filled with SS, the lower temperature product comes in contact with SS causing partial condensation, which transfers large latent heat content to feedstock resulting in a rapid rise in drying material temperature in addition to a layer of condensation of steam in close vicinity. This quick increase in the feedstock temperature is due to the release of the latent energy of condensation of steam, which itself is significant. Additionally, pre-condensation (high water activity) effectively helps in an inactivation of almost all microorganisms as compared to HAD as feedstock temperature increases much slowly and no moisture pre-condensation. However, a slower rise in the feedstock temperature will multiply the microorganisms, or they may generate toxins prior to inactivation.
iii. SS as the drying medium moreover permits other essential operations (blanching and parboiling, as well as puffing or popping) to be concurrently performed along with drying for some specific products.
iv. Typically, the feedstock dried in the SSD has a higher porosity than that dried in the HAD since water boils within the substance during SSD. In the feedstock, there is the vaporization (boiling) of water due to higher heating of SS, but if this generated steam cannot able to diffuse out at a certain rate, then the pressure will build up quickly, having a substantial volume expansion of the microstructure leading to have a high porosity of the dried material. This property is beneficial for a dried product, "any instant food" category, as it reconstitutes the dried product to its original form by absorption of water if it is immersed.
v. It is claimed that SSD improves the grindability index, reducing the power consumption for milling. At higher degrees of superheat and elongated exposure times, a reduction in the sulfur content of coal is reported in the literature; however, thorough investigation is desirable.

### 1.4.3 Limitations of the SSD

As the SSD is in the developing stages and new additions of technological and product-related knowledge explore the potential of this innovative technology, today's limitations may be overcome in the near future. The present limitations for SSD, which need to be understood for better applications, are listed as follows:

i. As SSD needs a steam generator as well as a superheater compared to a hot air generator for HAD, the former becomes a huge system, challenging to manage easily. Further, to take advantage of energy efficiency, heat recovery will be essential, which increases complexity and operational challenges and thus is high capital-intensive.
ii. Feedstock feeding and extraction of dried product from a drying chamber in SSDs must not allow infiltration of air or exfiltration of steam, which may increase energy loss and hence the operating cost. Therefore, SSD needs exceptional sealing at entry and exit of feedstock.
iii. The starting and stopping of the drying process plant are more difficult operations for SSD than HAD due to the involvements of the multiple systems, so special attention to skilled manpower is a must.
iv. At the beginning of the drying process, SS comes in contact with ambient feed, resulting in condensation of steam on its surface, increasing the quantity of moisture to be removed. This results in an increase of drying time by 10%–15%. In certain food product drying, it improves quality after drying of the product.
v. Heat-sensitive products require material temperature not more than 60°C–70°C, though it is subjective; because of that, specially designed LPSSD is more suitable, but it will further add in the cost to maintain the vacuum in addition to steam systems.
vi. Some substances may require oxygen during the drying for oxidation (e.g., enzymatic constituents such as vegetables, fruits, seafood, etc.) to generate preferred features such as color, taste, etc., which may not be feasible for SSD excluding non-enzymatic products. On the other hand, in this case, a hybrid approach of drying typically, steam drying along with HAD, may be suitable and it will give superior results such as energy efficiency and better product quality.
vii. Recirculation of the SS is not feasible or excess steam available in the dryer does not have any applications in the plant or nearby processes, then the saving of energy due to recovery of steam is no longer advantageous in SSD. Also, cleaning of recirculated steam is always challenging. Moreover, the quality of the condensate and contamination needs to be investigated appropriately so that it will not have adverse effects.
viii. Various peripherals, such as feeding equipment, dried substance extraction and collection, heat recovery and recirculation systems, are costlier than steam dryer alone. Therefore, for each application, thorough techno-economics of these auxiliary systems needs to be assessed.

The technological developments are still in progress and till date only a few products are commercialized having limited field experience with SSD. Over a period of time, this database is expected to increase significantly, as many more products are suggested for pilot testing especially for SSD.

## 1.5 APPLICATIONS OF SSD

The SSD has wide applications due to its unique features, but till date most of them are either at the laboratory stage or at pilot plants, and very few are fully commercialized. The important applications are summarized in Table 1.2 and also elaborated as follows:

### 1.5.1 Bio-Nonfood

i. Sludge drying is fully commercialized with capacity in tons for various types of dryers as fluid bed, flash, and dryers with agitated trough operated in SS medium.
ii. SS fluidized bed drying with internal heat exchanger tubes immersed within it for drying coal is a fully developed product. To improve the performance of SSD, exhaust steam from the first dryer can be used next, thus

**TABLE 1.2**
**Summary of Various Applications of SSD**

| Industrial Sector | Products |
|---|---|
| Biomass (for thermal power plant) | Pulverized coal, sawdust, wood fiber, wood chips, milled peat, wood pulp, peat for briquettes, spruce and birch bark, hog fuel |
| Chemical process plants | Powdery chemicals, washing powder, potassium salt, pigment, catalysts, epoxy (curing and drying), coloring agents |
| Food processing industry | Sugar: sugar beet pulp, dietary fiber from sugar beat |
| Spices, medicinal plants, herbs | Coriender and pepper seeds, Indian gooseberry, baccaurea pubera, Labiatae herbs: rosemary, peppermint, thyme, oregano, sage, spearmint and marjoram, korean traditional actinidia (Actinidia arguta) leaves, perilla oil seeds |
| Food drying | Bamboo shoot, cacao beans, herbs, paddy/rice, tobacco, corn gluten |
| Food drying value addition (source materials) | Corn fibers, protein-containing base substances (rape seed, soy beans, sunflower seeds, okara, soy meal, etc.), durian chips, yellow pea |
| | Vegetables: Potato chips, pre-drying, carrots, cabbage, onion slice, white radish discs |
| | Fruits: Mangosteen rind, ripe mango, apple cubes, orange powder, cashew apple (testa), banana slices, longan without stone, and asam gelugor |
| Food industry (starch) | Starch, distillers, grain wheat flour |
| Milk and milk processing (dairy plant) | Milk powder (flash drying), spray drying, paneer |
| Food industry (meat) | Shrimp, fish, zousoon, chicken meat, fish press cake; bone meal, pork, poultry food, fish meal |
| Paper and pulp uses | Drying Paper, fiber sludge extracted from wastewater treatment |
| Textile applications | Drying of coating line textile, heat setting and curing during fabric drying |
| Waste materials | Filter cakes, sludges generated from Industrial and municipal solid waste and effluent treatment, water treatment |
| Building materials | Wood, fibers, mineral wool, lime mud (before calcination), cement curing |

cascading alike multi-effect evaporators, generally in use for a concentration of liquids, may achieve economy of steam 1.9 for three-stage cascading (triple-effect SSD). Further, continuous fluid bed dryer and rotary dryer are also successfully tested for this application.

iii. Drying of lumber in the Moldrup process marketed by WTT (Wood Treatment Technology) of Denmark (formerly Iwotech Limited) is accomplished in the enclosed autoclave at vacuum drying with SS at below atmospheric pressure (10 kPa).
iv. Peat drying in SS flash dryers, Peco dryers, and tubular steam dryers have been used commercially. The pilot test of SS fluidized bed drying of peat at Helsinki University of Technology, Finland, is satisfactorily presented for a water evaporation capacity of 100 kg/h.
v. Drying of pulp in SS at 5 bar for flash dryer is used commercially and this dryer may have applications for drying sawdust, peat, and forest biomass, and, after pulverization, they may be useful as fuel replacing oil in existing thermal power plants.
vi. Though there are laboratory trials showing that the SS drying of tissues and papers yields superior quality, due to certain engineering and operational issues dryers are not yet commercialized fully. There need to be pilot studies for continuous operations mainly for impingement SS dryers.
vii. Wood wafers and particles drying using pressurized SS dryer pilot tests are completed successfully. Their fully commercialized systems are available in market.
viii. GEA Niro from Denmark had developed an industrial-scale pressurized superheated steam fluid bed dryer (SSFBD) for particulate materials and slurry/pulpy substances such as hey, brewer's spent grain (BSG) or draff, alfalfa, fish meal, waste of fruits such as peels and pulp from citrus, apple and pomace, bark, bagasse from sugar cane, wood chips, and municipal sewage sludge.

### 1.5.2 Bio-food

i. Drying beet pulp by SS with horizontal pressure vessel is manufactured by BMA AG of Germany. Other industry, GEA Niro from Denmark commercialized a pressurized SSFBD [30] for beet pulp drying.
ii. Soya sauce cake as feedstock processed in an agitated trough steam dryer for dewatering and deodorization is economic and good value addition [25]. Concurrently, SSD was successfully applied by Akao and Aonuma [31] for deodorization of fish meal, rice bran, and only drying with good quality of green tea, vegetables and silk cocoons.
iii. The impinging jets SS drying for potato chips was innovated in Japan, which gives better product quality (preservation of color and vitamin C).
iv. Paddy drying: A fluidized bed superheated steam drying (FBSSD) for paddy gives higher head rice yield and more sustainable.
v. Tortilla chips: It dries faster with impingement drying compared to HAD under the same conditions.

vi. Pilot- and/or laboratory-scale studies on coriander and pepper seeds in FBSSD, basil leaves in LPSSD, and SSD in edible Chitosan film were reported.
vii. Fish press cake with the impingement cylindrical dryer gives better product quality.
viii. Hybrid drying of banana slices in LPSSD and far-infrared radiation (LPSSD+FIR) had given acceptable product quality including porous structure compared to conventional vacuum with FIR.
ix. Carrot in LPSSD at 60°C was the optimal operating condition for the preservation of P-carotene and antioxidant activities as well as shrinkage, color, and rehydration characteristics with SSD.
x. Soybean drying using fluidized bed dryer is commercially available as an alternate to air drying.

### 1.5.3 Chemical and Other Materials

Dry spinning of synthetic fibers in superheated solvent vapor have advantage of stronger and finer fiber without surface wrinkles. Ceramic drying using SS is also developed having advantages of lower drying time and energy without much rejections.

## 1.6 IMPORTANCE OF SUPERHEATED STEAM DRYING

There are a total of 2297 papers found using the Scopus database from 1976 to 2023 with the keywords "superheated steam drying," "airless drying," "superheated steam drying with energy-saving," and "steam drying." The number of research papers published with the aforementioned keywords and cited in the Scopus database between 1976 and 2023 is shown in Figure 1.3. This number excludes master's and doctoral theses as well as unpublished industry work. The research papers are compiled from journals, conferences, books, book chapters, conference reviews, editorials, reports, and other publications. It indicates that, till 2000, the research on SS drying as media was very limited, but afterward it increased at a higher rate. Industrial activity in SS and closely allied fields can be assessed by examining the number of patents issues globally. The Espacenet Patent search engine was used to obtain the number of patents published annually around the globe. We used the keyword matching all words first "dryer," "drying" and then refined the search with "superheated steam drying" as well as "superheated steam." The total number of patents published each year over the period (1881–2024) were searched and are displayed in Figure 1.4. It is clear from these data that there is significant and continuing industrial R and D activity in developing SS technologies. Similar observations are noted in patents published as shown in Figure 1.4. However, it is still not sufficient to explore the full potential of this green technology for drying having other advantages explained in Chapter 2. Overall, a good amount of research work is in progress at laboratory scale, which needs to be transferred to industries by pilot- and industrial-scale studies to get the benefits of this emerging green technology.

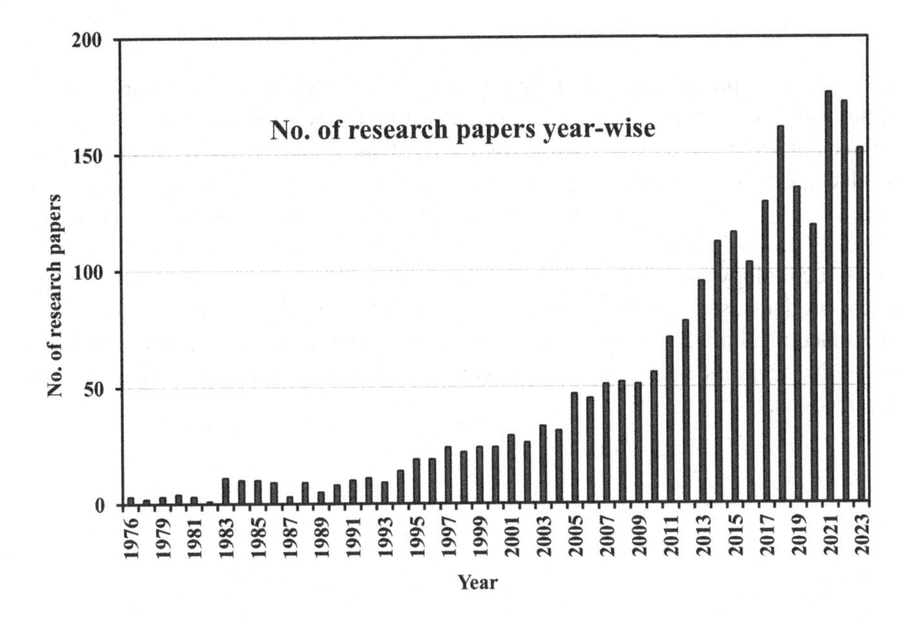


**FIGURE 1.3** Number of research publications with years [32].

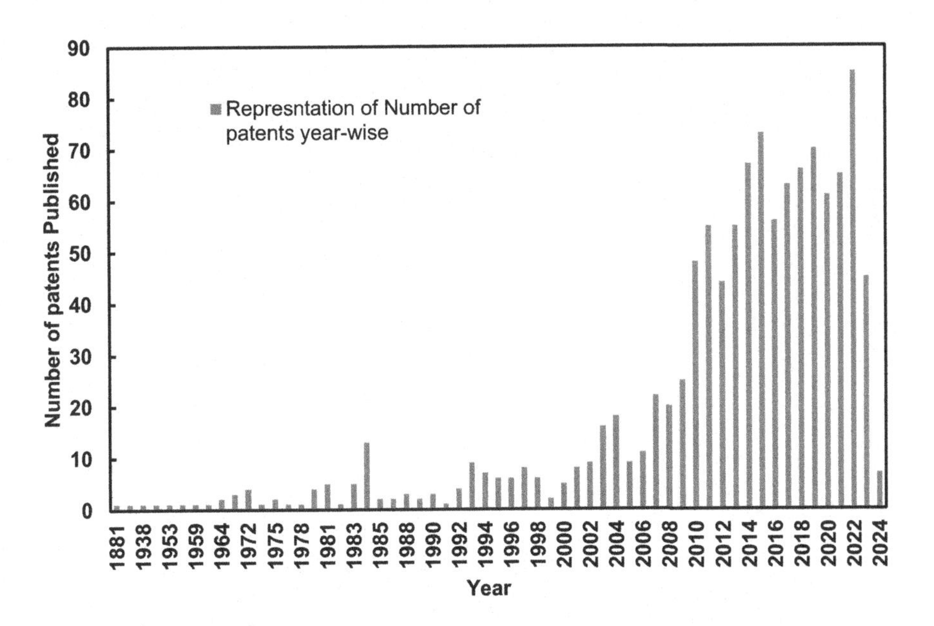


**FIGURE 1.4** Number of patents issued worldwide annually in the field of superheated steam drying and allied areas from Espacenet Patent search [33].

## 1.7 CLOSURE

SS drying is though more than century-old technology innovations on its wide applications are from the last three decades for diverse types of products due to unique advantages. It can be used at high- as well as low-temperature (heat-sensitive products) drying applications, but for the latter it is working at below atmospheric pressure. Further, it is minimal or no oxygen media being SS environment having typically useful in drying applications where oxidation reaction is undesirable. This green technology having lower energy and resources footprint can be used for different products, such as raw food preservation and processing, industrial chemicals, pharmaceutical product formation, drying of biomass, papers, and textiles. In recent years, its scope is increasing and moving toward the commercialization of wide applications.

# 2 Basics of Superheated Steam Drying

## 2.1 INTRODUCTION

Superheated steam drying (SSD), though reported almost 125 years ago active ongoing research from the last five decades, and recently, it attracted the attention of many more researchers. The foremost advantages of SSD reported by many researchers are energy-efficient technology, higher drying rate (minimum drying time), and better-quality products than hot air drying (HAD) with other significant benefits. It is differentiated from HAD based on the physical mechanism of a diffusion process for moisture transfer which is absent in it. This chapter deals with the working principle, layout, important components of the SSD, classification of the SSD, the significance of the SSD for reducing energy, carbon, and water footprints, and comparison between the drying media as hot air (HA) and superheated steam (SS).

## 2.2 OPERATIONAL PRINCIPLE

SSD is principally working similarly to HAD, in which the feed material is introduced to the SS environment to remove moisture content in it by convective heating. The representative Figure 2.1 of SSD depicts the foundation of operation with main components. At the start of the dryer, the steam generator produces the saturated steam at the required temperature, pressure, and quantity, which is passed to the superheater to provide required heat so it is superheated up to supply temperature at constant pressure. Eventually, it is appropriately supplied to the drying chamber to transfer thermal energy to dewater the feedstock and to take out the moist vapors from the chamber by the appropriate mechanism. Ideally, once the drying cycle starts, the partial exhaust steam is to be heated in superheater and saturated steam from the steam generator is no more required. Normally, in SSD, the outlet mixture of vapor (a combination of supplied SS, moisture evaporated, and volatiles if evolved) is partly recirculated into the chamber after heated to supply temperature into the superheater. Further, excess mixture of vapors contains significant thermal energy, which may be utilized judiciously in the plant itself for process heating, or its energy may be recovered. In case, the recirculation of exhaust gases is not compatible, then the whole exhaust mixture of vapor will be utilized such that there will not be any energy wastage. This would definitely lower the required net energy for drying. As discussed above, theoretically, the steam generator is operational during the beginning of the drying plant as exhaust contains surplus steam by the amount generated

DOI: 10.1201/9781003275299-2

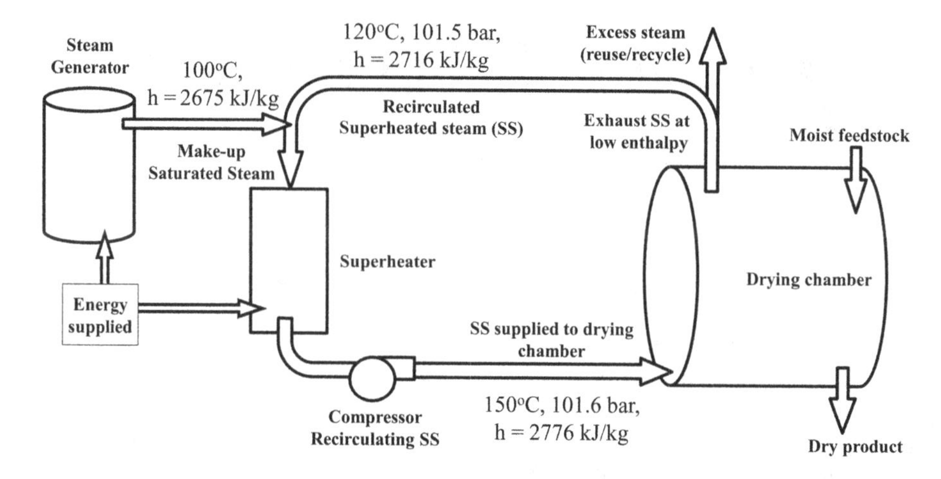


**FIGURE 2.1** Principle of superheated steam drying with steam generator.

due to drying and only the required quantity is recirculated back to the chamber after passing through the superheater. In practice, around 60%–65% of steam can be recovered, which is still significant [26,28].

In SSD, the heat transfer process between the drying medium and feedstock is having a higher rate and is more effective than HAD due to multiple reasons as stated above: (i) SS has a higher heat capacity and thermal conductivity than HA; (ii) both the drying medium and moisture to be removed have almost no resistance, resulting in a higher heat transfer coefficient; (iii) steam being water, it intensely penetrates wet products, but this is perfectly applicable for porous materials, reducing drying time by increasing the drying rate [28].

Since the steam density is lower than air, conveying is mainly constrained by proper handling of the dried material as SS is compatible with wide conveying systems.

## 2.3 COMPONENTS OF SSD

The SSD system is the combination of several main and ancillary components. The main components of SSD are a drying chamber, steam generator, and superheater. In contrast, the ancillary components used in this system are piping, valves, fittings and flanges, and a pump/circulation blower mainly applied for distribution, meter, or control the flow of steam and for feeding the feedstock from its point of generation to storage or supplying into the main components of a system.

### 2.3.1 Drying Chamber

The drying chamber is a prime and essential component of the drying system of SS dryers. It is used as enclosures for the feed/material (of which moisture to be removed) and drying media such as SS. A convenient drying chamber is designed for efficient transfer of moisture from the feedstock. The energy loss from the drying chamber is reduced to a minimum possible by appropriate insulation. Further, it provides the appropriate passage for the feed entry and product outlet based on the type of feed and dried product.

The structure and dimensions (shape and sizes) of drying chambers may vary significantly based on the applications of the dryers as experimental, pilot scale, or commercial. The schematic diagrams of the different shapes of the drying chamber for diverse types of dryers are presented in Figure 2.2. The capacity of the drying chamber can be as smaller as 5–10L, while the largest can hold several hundred liters.

### 2.3.2 Steam Generator

It transfers thermal energy evolved during the combustion of fuel safely, reliably, and efficiently to vaporize water and gets converted into saturated steam at the desired pressure and temperature. A simple layout is shown in Figure 2.3, where the boiler utilizes both sensible heat and latent heat to convert water into saturated steam. Superheater increases the temperature of the steam further, above the saturated temperature discussed as below.

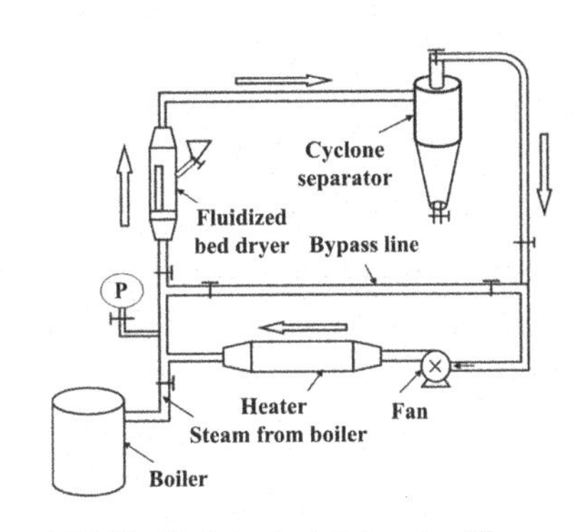


(a) Fluidized bed superheated steam dryer[34]

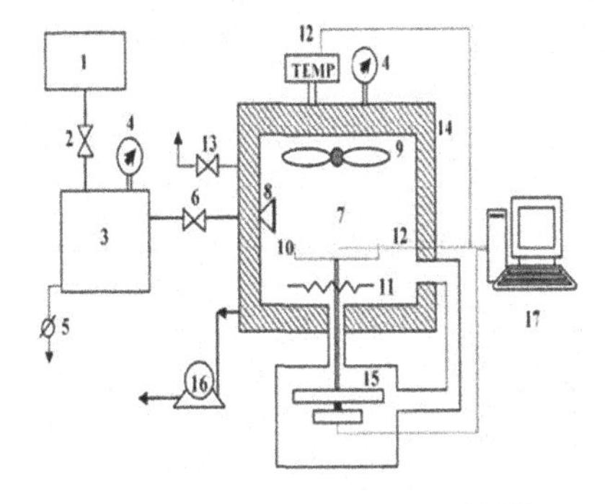


(b) Experimental setup of low-pressure SSD[35]

For Figure (b) (1) Boiler, (2) Steam valve, (3) Steam reservoir, (4) Pressure gage, (5) Steam trap, (6) Steam regulator, (7) LPSSD chamber, (8) Steam inlet, (9) Electric fan, (10) Sample holder, (11) Electric heater, (12) Temperature sensor and recorder, (13) Vacuum break-up valve, (14) Insulator, (15) Load cell and mass recorder, (16) Vacuum pump, (17) PC and data acquisition system.

**FIGURE 2.2** Schematic diagram of different shapes of drying chamber for various types of SSD.

*(Continued)*

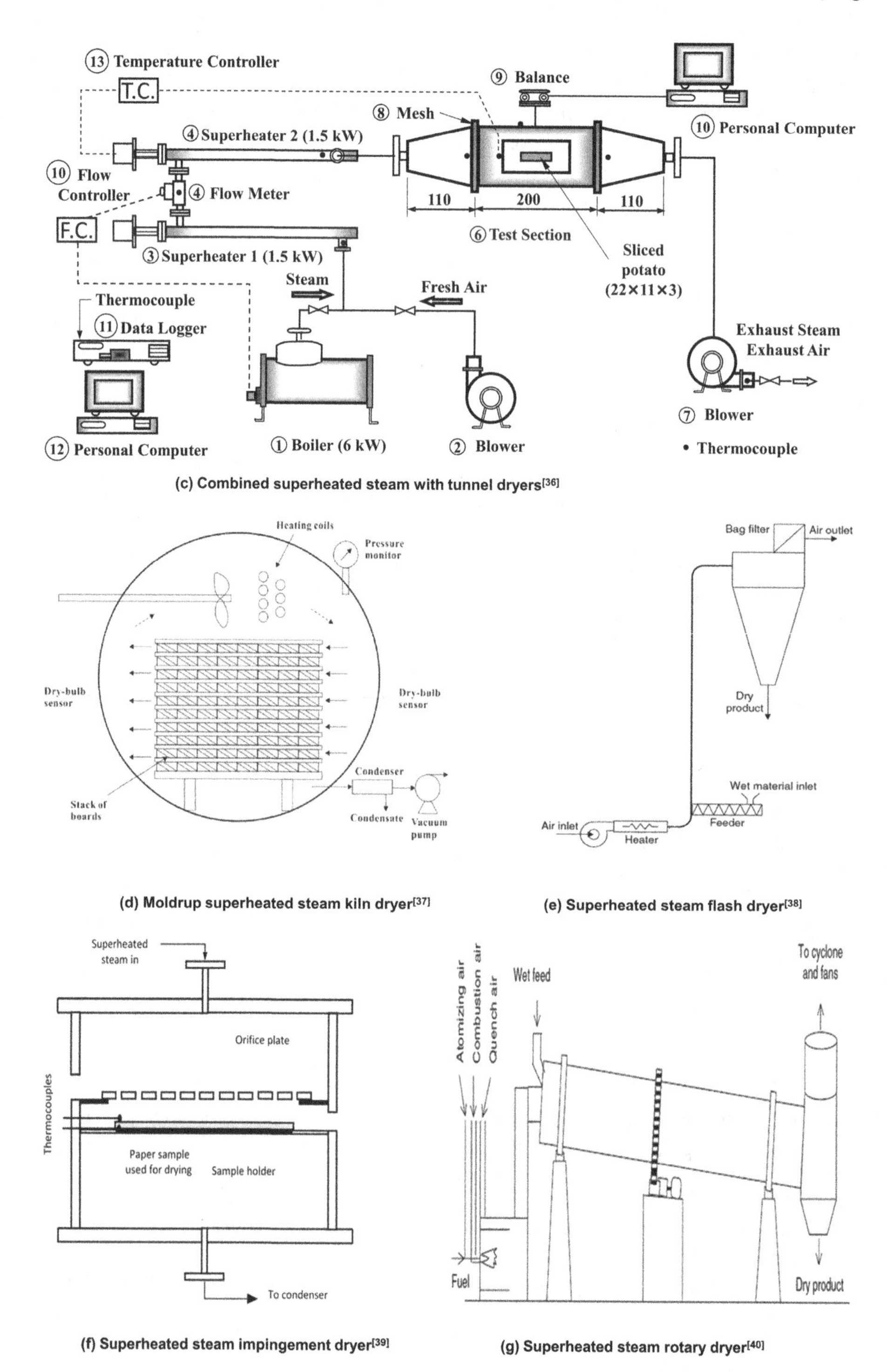


(c) Combined superheated steam with tunnel dryers[36]

(d) Moldrup superheated steam kiln dryer[37]

(e) Superheated steam flash dryer[38]

(f) Superheated steam impingement dryer[39]

(g) Superheated steam rotary dryer[40]

**FIGURE 2.2 (*Continued*)** Schematic diagram of different shapes of drying chamber for various types of SSD.

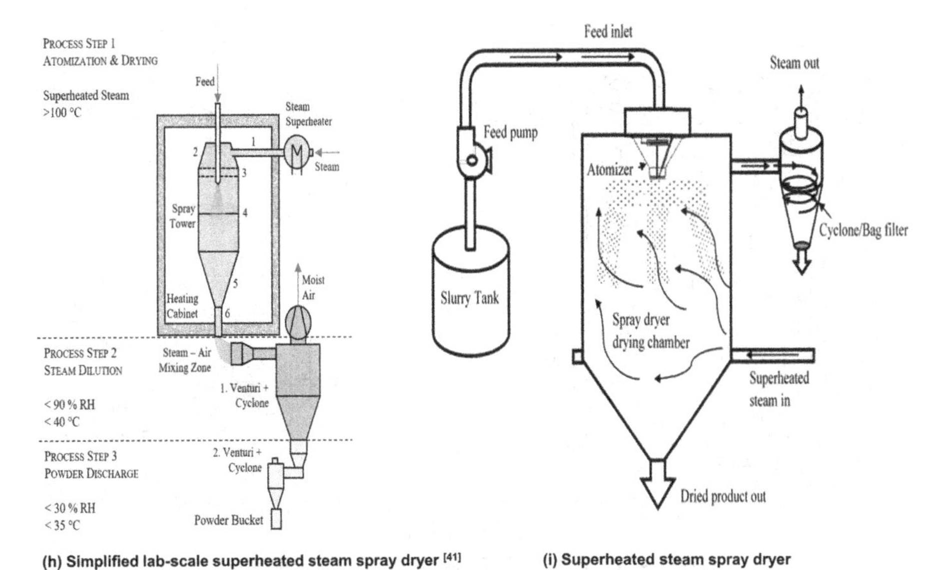


**(h) Simplified lab-scale superheated steam spray dryer [41]** **(i) Superheated steam spray dryer**

**For Figure (h) (1) superheated, steam inlet, (2) tower lid, (3) gas distributor, (4) cylindrical segments, (5) conical segment and (6) discharge pipe.**

**FIGURE 2.2 (*Continued*)** Schematic diagram of different shapes of drying chamber for various types of SSD.

### 2.3.3 Superheater

In the superheater, superheating of the steam was performed by separate equipment providing additional heat transfer surface downstream of the boiler as indicated in Figure 2.3. It is typically a heat exchanger utilized for the generation of SS from saturated or in some systems, it converts the wet steam generated by a boiler to dry steam and then subsequently to superheated form appropriately using the thermal energy of hot flue gases/air. Superheater utilizes sensible energy for transferring saturated steam to superheat form by increasing its enthalpy.

In case of SS dryer with recirculation of exhaust steam, a separately fired superheater is available due to continuous operation but the steam generator needs only during the starting of the drying plant. Superheaters are mainly of two types: (i) coil type and (ii) shell type. Coil-type superheaters usually contain several coils welded on both ends to the headers. Generally, steam passes through the coil, and flue gases are surrounding the coils. Shell-type superheater is typically used to superheat the steam electrically by the distinctive heating system stacked in a shell passing the flow of steam over it. Moreover, based on the required steam temperature, the superheater is classified as: (i) radiant superheater for very high temperature, so not preferred in the SSD, and (ii) convective superheater for moderate temperature, thus preferred for SSD.

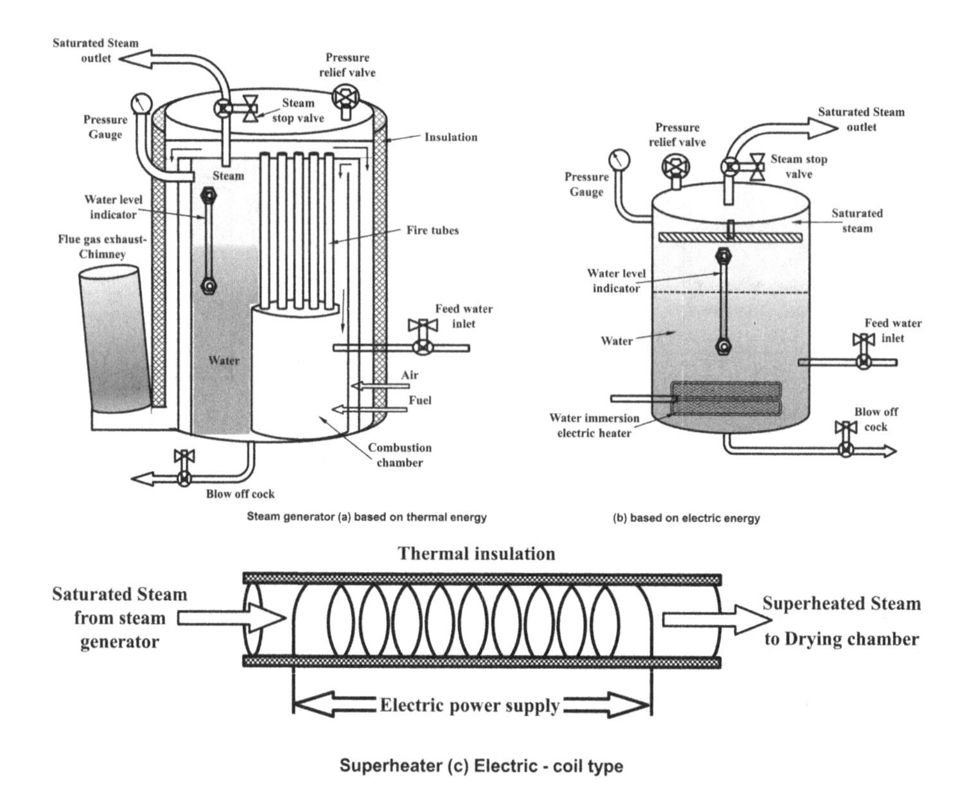


**FIGURE 2.3** Schematic diagram of steam generator and superheater. Steam generator (a) based on thermal energy and (b) based on electric energy. Superheater (c) electric-coil type.

## 2.4 SUPERHEATED STEAM THERMOPHYSICAL PROPERTIES

The thermophysical and chemical properties of any drying medium such as specific heat capacity at constant pressure (Cp), density, thermal conductivity, viscosity, relative humidity, and wet bulb temperatures are very important for deciding the temperature of the feedstock, heat transfer rate, moisture transfer, and drying rate and time. For the case of the SS, Cp, density, thermal conductivity, relative humidity, and wet bulb temperatures are having greater values than HA/gases at atmospheric pressure in the operating temperature range of the dryers (around 150°C–180°C) [1]. Further, SS and evaporated moisture being same phases of water, the diffusion resistance for moisture transfer is practically almost negligible. Viscosity, one more important property, is lower for SS than HA at the same temperature range, indicating lower resistance and favorable for the drying process. However, it may reduce residence time, which can be compensated by the longer length of the dryer or increasing distance traveled by SS. All these lead to a higher heat transfer coefficient for SS compared to HA. However, the temperature difference between the feedstock and the drying media (SS) during the constant rate period is lower than HA/gases since the wet bulb temperature is pretty high for SS (100°C at 1 atmospheric pressure). Overall, these two parameters, heat transfer coefficient and temperature driving potential, lead to the so-called inversion point temperature at which the evaporation rate has the same value for both the mediums (HA and SS) as discussed in detail in Section 2.11. For SS compared to HA, the greater heat capacity and convective heat transfer coefficient leading to higher heat transfer are also beneficial for decontamination of the microorganisms. This inactivation of microorganisms is leading to a hygienization of dried products. Degradation of the food stuffs to be dried also depends on the exposure time to drying media and, in the case of SSD, a higher drying rate culminates in a short retention time, increasing degradation of it. However, in this particular case, exposure time will be designed so that the product life will not affect.

## 2.5 CLASSIFICATION OF SSD

The SS dryers are having wide applications in various dryers reported by numerous researchers. For systematic study and else of selections of SSD, classification depends on the multiple criteria as discussed in the literature and displayed in Figure 2.4 and deliberated in the following paragraph [1,7]:

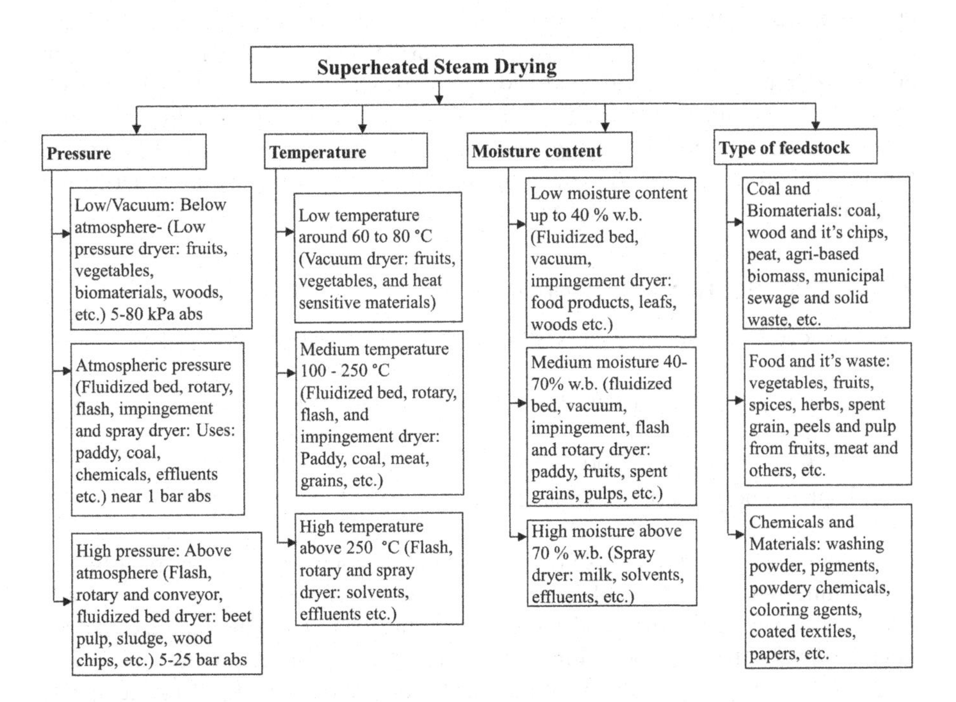


**FIGURE 2.4** Classification of superheated steam drying based on various criteria [3,7,29].

### 2.5.1 Dryer Operating Pressure

The SS dryers are grouped into three categories, viz.:

a. Low/vacuum pressure (pressure: below atmosphere)
b. Atmospheric pressure (pressure: little higher than atmosphere)
c. High pressure (pressure: quite higher than atmosphere)

Low-/vacuum-pressure superheated steam drying (LPSSD) is selected especially for substances which are heat-sensitive like food: vegetables, fruits, spices, herbs; bio-products: lumber, grass, biomass. The required drying temperature for feedstock is between 50°C and 80°C (in constant drying region), so to maintain it the saturation temperature of the SS should be around the same range, which is possible by lowering the pressure below the atmosphere. An initial idea of LPSSD was put forward by Mujumdar [16] in the early 80s for silk drying. The pioneer work for the food drying at low pressure, such as carrots, cabbage leaves, Indian gooseberry, banana, etc., is reported by Devahastin et al. [35] and Devahastin and Mujumdar [26].

For the next category, i.e., atmospheric-pressure SSD, the drying takes place at 100°C in a constant rate drying region. After that, product temperature will be increased maximum up to SS temperature. This is typically the most studied SS dryer and commonly used for various applications such as tissues, paper, ceramics, industrial effluents, coal, detergents, certain food and other items, dairy products, etc., compared to the remaining two types, low- and high-pressure SSD. The last category, high-pressure SSD, is applied for the drying of sludges, biomass, beet pulp, etc., with dryers as flash, fluidized bed, rotary drum, and conveyor, and typically works at above atmosphere (5–25 bar ab). Normally, the use of SS at high pressure (more than atmosphere) for drying applications has no substantial effect on drying kinetics. Figure 2.5 shows the operating regions of low-pressure, atmospheric, and high-pressure SSD with reference to saturated steam.

### 2.5.2 Superheated Steam Temperature

SSD is categorized into three classes based on the SS temperature as follows:

a. Low temperature (lower than 100°C, similar to low-/vacuum-pressure drying)
b. Medium temperature (100°C–250°C)
c. High temperature (higher than 250°C)

In certain cases, it may be possible to operate the low-pressure SS dryers at higher temperature (above 100°C). The variation of steam temperature is having considerable influence on the time and rate of drying as well as product quality, mainly microstructures than the above atmospheric pressure. The rise in pressure of SS will only change the saturation temperature and do not influence the feedstock temperature at least in constant drying rate region. Furthermore, the sensible energy content of steam (energy in the superheated region of the steam) is merely employed for the dewatering of the feed, keeping pressure almost constant (no pressure drop), and

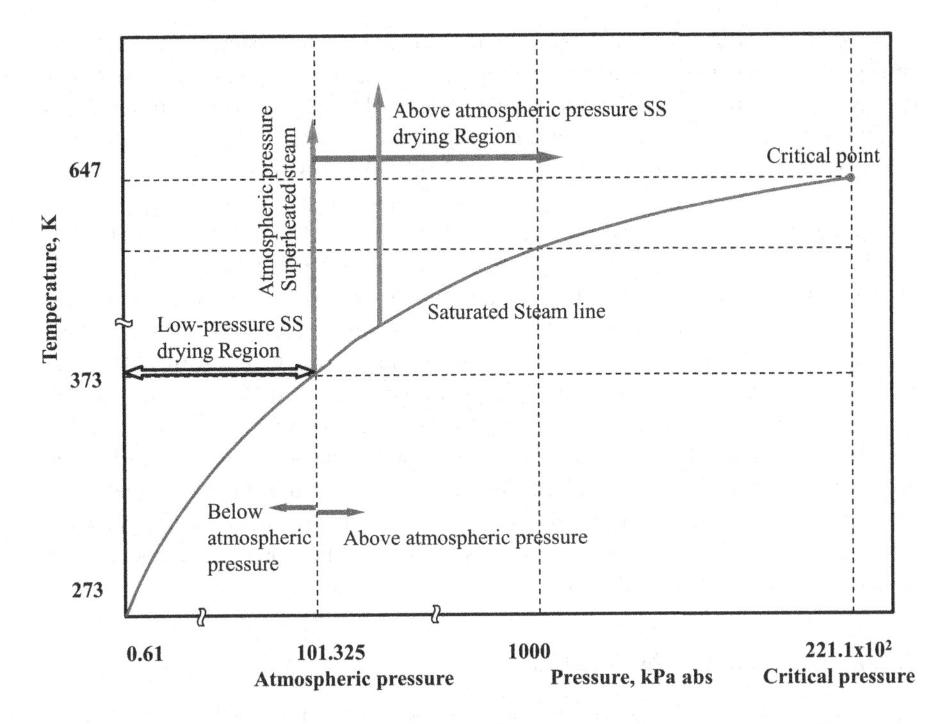


**FIGURE 2.5** Representation of SSD at various pressure and temperature conditions (not to scale).

latent heat is not used for the same as energy source. The energy recovery from exhaust steam has positive effect in case of pressurized dryer.

As discussed above, the temperature of the SS (a drying media) has primary influence on thermophysical properties of the steam, which are driving the drying phenomena. The SS temperature less than 100°C is typically possible for chamber pressure lower than atmosphere and is considered low-pressure SSD. It is essentially applicable for heat-sensitive materials such as fruits and vegetables, dairy products, etc. The SSDs operating in the range of 100°C–250°C are classified as medium temperature dryers and are most extensively used commonly at atmospheric pressure. Further, there are favorable drying kinetics and distinctive characteristics, because the inversion temperatures for most of the feedstocks are around 160°C–170°C for the SSD media.

### 2.5.3 Moisture Content in the Feed

Moisture in the feedstock varies from substance to substance and based on that, the energy requirement is determined. Considering this, it is divided into three categories, as given below:

a. Low moisture content (less than 40% w.b.)
b. Intermediate moisture content (40%–70% w.b.)
c. High moisture content (higher than 70% w.b.)

The feedstock with low and intermediate moisture content categories are herbs, spices, food products, shrimps, beverages, sludge, woods, coal, etc. Being an energy-efficient drying technology, SSD will be used for drying these types of product, typically using low-pressure or atmospheric fluidized bed, impingement dryers, flash, etc. Additionally, highly moist items, mainly liquidus such as solvents, juices, and effluents, may be dehydrated by the spray dryers. Most commonly, spray dryers are applied to get powder/granular dried products from the pumpable feed. The drying time depends on the diverse parameters such as an internal structure of feedstock, moisture migration mechanism from it, and the influence of drying rate on product quality, and not specifically on the quantity of water in the feedstock before drying.

The next category is based on the type of the feedstock, which is further classified as: (i) coal and biomaterials in which materials such as different types of coal, wood, and their various forms such as chips, pulp, sawdust, etc., agri-based biomass (rice husk, sugarcane bagasse, etc.), municipal sewage and solid waste, etc.; (ii) food and its waste: vegetables, fruits, spices, herbs, peels and pulp from fruits, meat, bone meal, etc.; (iii) chemicals and materials: pigments, potassium salts, washing powder, powdery chemicals, coloring agents, catalysts, papers, lime mud (before calcination), coated textiles, mineral wool, and fibers.

### 2.5.4 Commercially Available Dryer

- Belt conveyor dryer: This dryer has been designed originally for food products. The unit can be opened and can be cleaned very easily. The belt conveying system allows the processing of large types of several products. The drying chamber is split into four different zones, as shown in Figures 2.6 and 2.7. The steam temperature and the flow velocity can be adjusted separately for each zones.
    - Evaporation capacity: 20–40 kg/h (depending on the product moisture content and steam temperature)

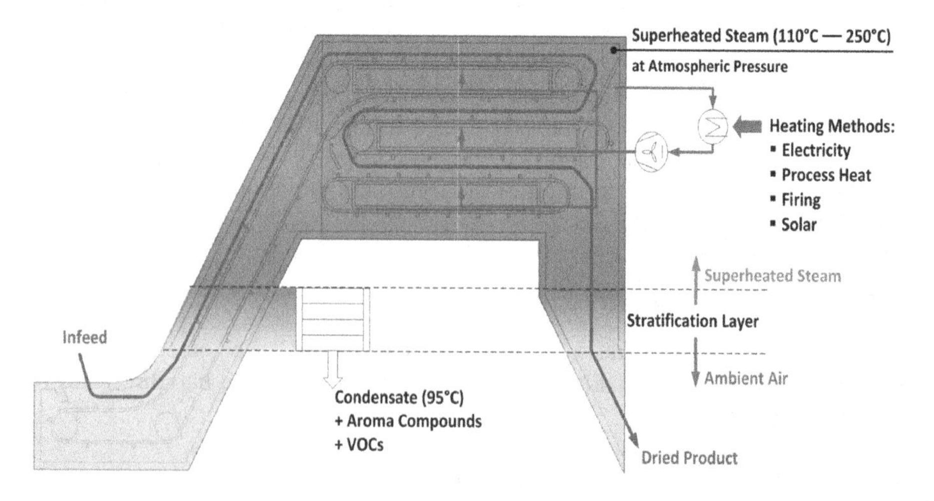


**FIGURE 2.6** Belt conveyer dryer. (Courtesy: Fraunhofer IGB.)

FIGURE 2.7 Photograph of conveyer belt dryer. (Courtesy: Fraunhofer IGB.)

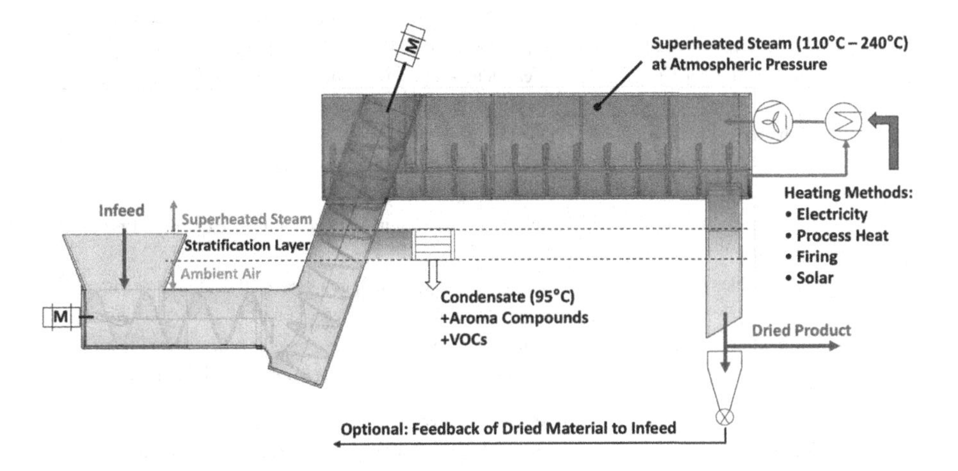


FIGURE 2.8 Spiral conveyer dryer. (Courtesy: Fraunhofer IGB.)

- Temperature max.: 200°C
- Spiral conveyor dryer: This dryer has been designed originally for pasteous and powdery products. Three shaftless spirals are used to convey the material (feeding spiral, main spiral, and exit spiral) as presented in Figure 2.8. Steam temperature and flow velocity can be adjusted. The exit spiral is equipped with a cooling system to avoid self-ignition or dust explosion at the product outlet.
- Evaporation capacity: 4–10 kg/h (depending on the product moisture content and steam temperature) (Figure 2.8)
- Temperature max.: 250°C

## 2.6 SELECTION OF DRYING MEDIUM

There are various aspects to selecting the appropriate drying media based on the priority. Table 2.1 gives succinct features of SSD, which are pivotal for the appropriate selection of the dryers based on process and product characteristics. Hence,

**TABLE 2.1**
**Important Parameters and Their Impact on Superheated Steam Drying**

| S.N. | Factor | Impact |
|---|---|---|
| | | **Product Related Factors** |
| 1 | Temperature sensitivity | As per requirement dryer can be selected as high pressure for higher temperature; atmospheric pressure- medium temperature; and to avoid higher steam temperature, vacuum system. |
| 2 | High moisture content | Energy-efficient dryers to reduce operating cost without affecting product quality. The latent energy of the excess moisture may be recovered in SSD with additional benefits of water recovery, improving the overall efficiency. |
| 3 | Pre-Post treatment of product to be dried | Sometimes, it may be possible to connect the pre or post treatment with a superheated steam media such as sterilization, pasteurization, blanching, etc. |
| 4 | High thermal resistance | Due to higher saturation temperature of SS, feed temperature is also higher which may decreases drying time due higher heating rate. |
| 5 | High sensitivity/low tolerance for oxygen | Absence of oxygen in SSD may improve quality of product. |
| 6 | Undesirable taste | As it is closed system, it can very well deal with products with rancorous and foul gases emitting during drying. |
| 7 | High product values | Drop in drying time have multiple benefits, like reduction in inventory and energy cost resulting in rise in product value. |
| | | **Process Related Factors** |
| 1 | Additional usefulness of surplus steam | The requirement of the energy use is less so the capital cost can be minimized by using the steam for other applications. |
| 2 | Environmental emission from dryers | SSD being closed system recovery of solvents and particulates will be done easily. |
| 3 | Explosion/combustion hazards | In SS environment, oxygen is hardly present (meager), considerably reduces fire/explosion hazards. |
| 4 | Expensive source of thermal energy | Thermal energy saving with SSD will offset greater energy costs than waste fuels (hogged wood waste). |

laboratory testing and pilot operation of the feedstock are required before continuing any decision to consider SSD.

## 2.7 SELECTION OF TYPE OF DRYING SYSTEMS

There are more than 500 different types of dryers and out of that, around 100 are commercially available. Each dryer operating with HA can be principally operated using SSD (e.g., spray, fluid bed, impinging jet, conveyor dryers, flash, etc.). However, technological conversion from HAD to SSD is more complex due to the involvement of more subsystems. Because of the unique features and structure of the products (product shape, size, internal molecular structure, etc.), and production capacity, the selection of appropriate dryers is crucial. Further, it is challenging to make a uniform procedure for the selection of the dryer, so important guidelines are provided here as follows:

i. The tray or shelf dryers are generally preferred for materials that are easily worked with trays.
ii. For the drying of temperature-sensitive feedstock's, low-pressure SS dryers are suitable as pressure below atmospherere (vacuum), the temperature of the feedstock in the range of 60°C–80°C. The water vapors generated will be condensed in a condenser positioned in between the vacuum pump and the drying chamber.
iii. Dewatering of particulate matters mainly in the granular or crystalline form, can be appropriate for rotary drum dryers. It comprises a slightly inclined rotating cylinder to the horizontal so that the feed entered from one side can be suitably delivered from another end. Energy will be provided by a row of concentric tubes attached inside the cylinder supplied with steam in addition to the SS drying medium, as discussed by Perry [42]. SS being airless closed system drying, solvents extraction and recovery, and drying of fire hazards materials can be safe and with else in it as noted by Perry [42].
iv. For materials which are light in weight or more-sticky and may not form bundles, agitated dryers are more preferred, which can also be dried by rotary dryers. The rotating scrapers continuously stir the feedstock to be dried from the heating surface of the jacket. These dryers are more suitable for the continuous process, despite that it is also accepted for batchwise processes. It can be feasible to use with low-pressure operation (vacuum) for the causes mentioned above.
v. For materials which are difficult to pass through the drying chamber may be due to high viscosity, such as semisolid or pasty; screw or paddle dryers may be useful and it can be heated from inside, transferring this energy to the drying material.
vi. The drum dryers are also suitable for liquidus or slurries, which are easily pliable between the rotating drums. It contains two cylindrical drums internally steam-heated, rotating like a roller, press in the direction of each other. Feedstock in the form of liquidus or slurry feeds into the portion between two rollers (V-type section). Doctor knives are provided to take out dried material from the outer surfaces of the drum. It may be restructured to suit low-pressure (vacuum) drying by an enclosure having a vacuum.
vii. The drying of textiles and papers like a continuous sheet (shaped appropriately) are passed in a zigzag manner over a series of cylindrical rollers rotating at the speed of the sheet produced. The extension of this dryer is a conveyor dryer made up of an endless belt of wire mesh suitable for pasty substances, on which these materials are passed.
viii. Particulate or granular material can easily form a particle flow and is typically competent for a fluidized bed dryer. The fluidized bed superheated steam dryer (FBSSD) can be operated purely on SS for fluidization and

additional heat can be supplied by a heat exchanger placed inside the fluidized bed, which further reduces SS flow and increases the efficiency of the drying system. Furthermore, vibrated fluid beds, agitated beds, and spouted and spout-fluidized beds are employed for specific applications taking advantage of their special features.

ix. For drying of suspensions, pastes, solutions, and slurries, spray dryers are by far the most popular direct dryers in use due to their ability to produce products in desirable form. Major applications exist in dairy (milk, whey, etc.); foods (coffee, chocolate, etc.), chemical, ceramics, detergent industries, etc. More recently, fluidized beds, vibrated beds, spouted beds, and impinging stream dryers utilizing beds of inert particles are competing successfully against spray dryers for certain smaller-scale applications not requiring the unique product characteristics only attained by spray drying. Combination of spray and fluid bed dryers has become more popular recently due to their reduced capital and energy requirements.

x. It should be noted that mixed-mode drying systems, such as flash and fluid bed, spray and fluid bed, mixed fluid bed followed by plug flow fluid bed, etc., have definite advantages that should be considered carefully in the selection of a proper drying system.

## 2.8 PHYSICS OF SSD

Similar to other traditional dryers, SSD comprises the transfer of the mass and thermal energy between feedstock and the SS as drying media. There are mainly two popular drying phenomena such as droplet drying of water with solid content and film drying over the solid. Latter will have different correlations for determination of heat transfer coefficient based on the orientation of the solid may be horizontal, vertical and inclined, and active sides as top, bottom or both in case of horizontal film drying.

Drying is constituted of more than two sub-processes segregated by the temperature and moisture content in feedstock or rate of drying as shown in Figure 2.9. The drying begins with, first, an initial heat-up period (induction period); in this period, the wet feedstock absorbs the thermal energy from its surrounding medium, heating it still the temperature above wet bulb temperature at which evaporation of water initiates from the feedstock surface. At the same time, there is pre-condensation of the SS, explained afterward in details. Second, it is generally represented as a constant drying rate period, in which solid surfaces will remain wetted with water till the end of it, thus drying rate and temperature of feedstock are constant during this period. Third, moisture in the solids will be reduced so much that there is no surface moisture but is inside the solid (bound moisture) may be through colloidal and capillary moisture, chemically attached moisture and physically adsorbed, respectively. In this region, a reduction in the moisture content in feedstock reduces the drying rate continuously with rise in required drying temperature. This region also observes the rise in the temperature of the feedstock. Overall characteristics of the drying in HA and SS are similar. Each drying region is explained in more details in the following sub-sections.

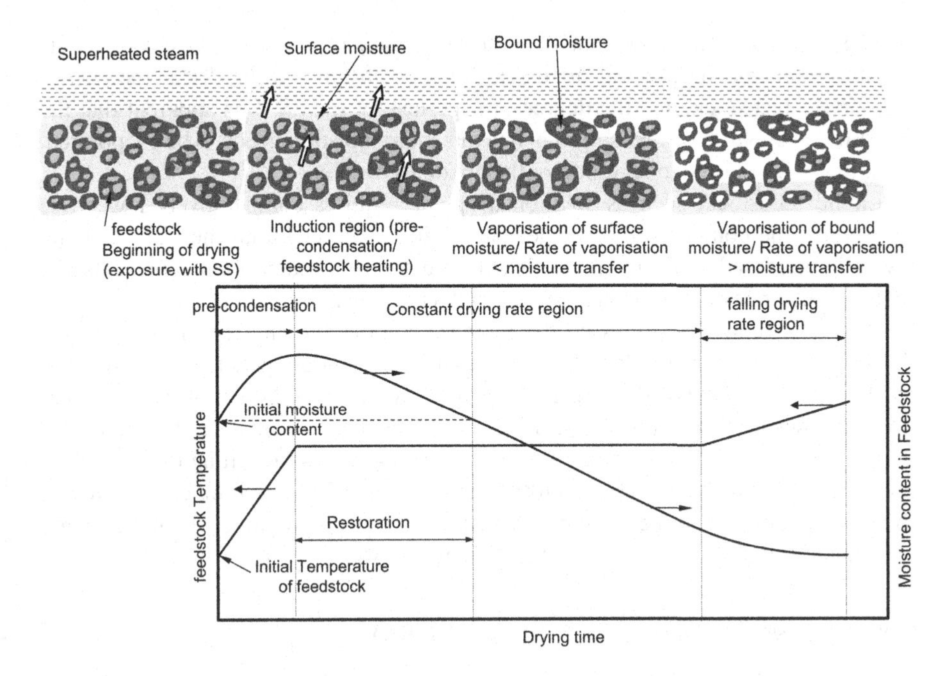


**FIGURE 2.9** Various drying regions in SSD.

### 2.8.1 Heat-Up (Warm-Up) Region

This region is divided into three sub-regions: (i) condensation, (ii) restoration, and (iii) drying region [43], as shown in Figures 2.9 and 2.10. Generally, feedstock at ambient conditions along with SS is supplied to the drying chamber, where heat transfer takes place from the SS to the feedstock. The temperature of the ambient feedstock starts increasing by reducing the temperature of the SS medium due to energy exchange in the drying chamber; this may lead to condensation of the part of the SS on the moist solid surface if the temperature falls below the saturation temperature, which is referred to as pre-condensation sub-region. This is observed for several feedstocks like spent grain pellets [44], distillers spent grain [45], *yellow pea* [5], etc. In this region, due to phase change (pre-condensation), the energy transferred at a high rate, resulting in a sudden rise in the temperature of the feedstock, which may be favorable in certain materials. After reaching a temperature of the feedstock same or a little above the saturation temperature of SS at chamber pressure, evaporation of moisture (drying) will begin from the surface of the solid. The restoration sub-region is a part of the region where pre-condensed moisture is evaporated and moisture content in the feedstock will be the same as the pre-drying state, i.e., restoration of pre-drying condition of the feedstock. The time required to remove the condensed moisture from the surface of the feedstock is referred to as restoration time, which increases drying time due to pre-condensation. Actual removal of moisture from the feedstock starts only after the restoration sub-region. Sometimes, the heat up region may be performed outside the drying chamber and hot feedstock will be entered into the drying chamber, where pre-condensation of SS will be avoided.

The quantity of initial condensation of the SS depends on several aspects like its moisture content, the thermal diffusivity of the substance to be dried, the energy capacity of feedstock, and the degree of superheat of the SS. It is apparent that the moisture level will be increasing. This means more moisture needs to be evaporated, resulting in more time to re-vaporize the condensed water.

Liquid feedstock, such as solution droplets, requires comparatively lower rise in drying time compared to solid feedstock due to lower initial moisture content. Overall, increased moisture in feedstock due to pre-condensation represents the higher moisture evaporation and increased drying time.

Various researchers discussed the increase in moisture and rise in drying time during heat-up regions for diverse materials such as cellulose pellets by Luikov [46], food by Yoshida and Hyodo [47], drying water droplets and droplets of several solutions and suspensions by Lee and Ryley [61], and fries, vegetables, herbs, wheat, flour, cacao nuts, etc. by Trommelen and Crosby [48]. After reaching the saturation temperature at chamber pressure (100°C at 1 atmospheric pressure), evaporation of the surface moisture of the feedstock will start (drying sub-region), which is referred to as the constant rate drying period explained below.

### 2.8.2 Constant Drying Rate Region (CDRR)

As mentioned above, the CDRR begins from the completion of the heat-up period and ends at a point of the time when the rate of the internal moisture transfer is not sufficient enough so that the surface of the feedstock can be maintained appropriately moist (wet). The feedstock moisture content at the point where the drying rate

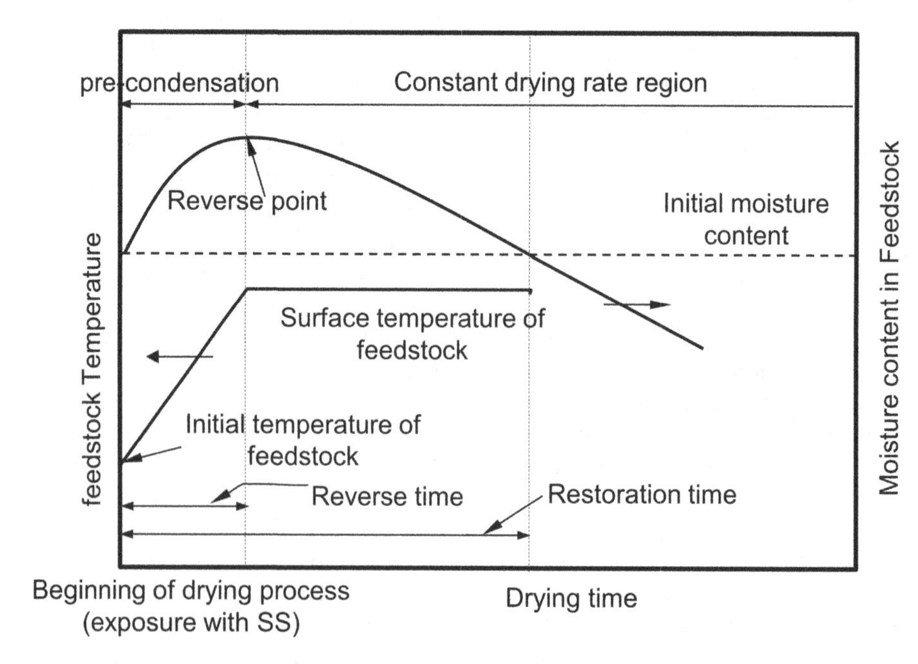


**FIGURE 2.10** Induction region of drying and its sub-regions.

just starts falling is called as critical moisture content and its value depends on the product features as well as the drying process. It is noticed during the experiments that critical moisture content for a particular feedstock (when the drying rate starts to fall) is generally lower in the case of steam drying due to enhanced mobility of the liquid and vapor moisture through the solid matrix. This enhances the drying process, which now proceeds over an extended moisture content range at a higher rate, thus reducing drying time and dryer size in some cases.

In this case, initially surface moisture (water layer present at surface) is evaporated due to thermal energy supplied by SS and then moisture inside the feedstock is transferred at higher rate than the rate of evaporation of moisture such that the solid surface is maintained wet. Because of this, solid surface temperature remains constant at saturation temperature corresponding to the chamber pressure [13,17,19,49]. For the moisture movements from the inside to surface of the solid, there are various mechanisms and are discussed in fundamental books of drying [1,26,28].

For the superheated vapor environment drying, thermal energy is transferred similar to air/gas drying via a vapor film to the surface of the feedstock to be dried under the effect of a driving potential of temperature difference. On the other side, the surface moisture removal is mainly because of the bulk movement due to driving potential of pressure difference while it is in SS environment and not diffusion mass transfer as conventional HAD. Chu et al. [11] observed that the increase in surface temperature by fraction of degree Celsius (only a few hundredths of a degree) offers a suitable pressure drop to have the evaporated moisture flow. Considering the case of 150°C SSD of 1 mm diameter water droplet, it requires pressure drop of only $10^{-6}$ N/m$^2$ between SS medium and drying surface in atmospheric pressure drying system for the sufficient moisture removal rate. In SSD, the moisture removal resistance from the surface water to become part of the vapor is so small, therefore it can be neglected practically.

For droplet drying, its surface temperature for a certain time period remains almost constant identical to the saturation temperature at that pressure. Since the droplet size diminishes as the drying proceeds due to evaporation of the water from it, drying rate does not remain constant (so cannot use the terminology as constant drying rate region); but during this period, the droplet surface temperature is constant equivalent to saturation temperature of moist surface and can be named constant temperature drying period having similar significance as that of constant drying rate period [48]. The drying rate during this depends upon the temperature difference between the bulk SS and droplet surface (feed), and appropriate heat transfer coefficient, but not on removal of vapor from the surface. Various researchers gave the correlations for determinations of the heat transfer coefficients.

### 2.8.3 Falling Drying Rate Region (FDRR)

As drying progresses further, the surface moisture of the feedstock/solid is vaporized completely, which results in the beginning of the fall of drying rate and hence the region from this point onward still drying completes (reaching equilibrium moisture condition) is called the falling drying rate region/period. In this region, the drying kinetics (exchange of moisture and thermal energy) of the feedstock is influenced the characteristics of the substances such as microstructure and the bonding of the water within it, which is unlike to constant drying rate period. High temperature of the material to be dried in SS environment results in higher moisture diffusivity in it as mentioned already than HAD especially in this region. Yoshida and Hyodo [47] compared fibers drying in air and superheated acetone environment, as well as foodstuff in superheated vapor, and noticed that vapor-dried materials produce porous and permeable surfaces due to the vapor flow within it, which makes superior products along with higher drying rate than HAD. Trommelen and Crosby [48] observed that film-forming substances like detergent and food products can have higher drying rate in SS than HA media in falling rate period due to mainly low resistance to transfer of vapor through the more pliable film that results on SSD.

Overall, drying rates and the general characteristics of the drying process are not much changed due to the drying media either superheated vapor or HA. Total drying rates (combined constant and falling drying rate) can be higher or lower in SS or HA reliant on several parameters such as the pre-condensation of SS in induction period, SS temperature compared to inversion temperature, the drying characteristic of the feedstock, etc. Prediction of heat transfer rates for SS medium can be appropriately adapted from correlations developed for drying using HA. One of the important changes required is the resistance to both vapor and moisture transfer within the feedstock seems to be lowered for SSD than HAD, rising the drying rate in the falling drying rate periods.

## 2.9 MODELING OF THE SUPERHEATED STEAM DRYING PROCESS

To understand the physics of the drying phenomena and manipulate it as per the requirement, modeling of the drying is important. Further, the models can be useful tool for simulation, which gives preliminary assessment of the technical and economic feasibility of the concept as well as in sizing and selecting the optimal operating parameters without expensive experimentation at the pilot or industrial scale. It is useful in minimizing but not eliminating the need for experiments. Product quality effects cannot, however, be simulated. Yet, they may govern the ultimate choice of dryer or drying system. There are basically two types: macro modeling and micro-modeling explained in the following section.

### 2.9.1 Macro Scale Modeling

Considering the overall system and its components as control volumes and using mass and energy balance equations across the boundaries of control volumes, the developed mathematical model is a macro scale model as it is accounting the system level parameters and not differential element. This model is useful to determine the rating of the dryer, sizing of the various components involved viz. drying chamber, superheater, circulation fan, ducts energy recovery systems, steam generator, etc., and parametric analysis to determine optimal operating condition. It can be further useful to compare SSD and HAD based on various criteria.

### 2.9.2 Micro Scale Modeling

In this small element of the physical system is analyzed for transfer of mass and energy using basic principles of science as well as property relations. This gives either differential or integral equations which can be solved appropriately by analytically or numerically. This model is very close to physics of the system and is able to predict at each space and time coordinate what is exactly happening in the drying phenomena. A more detailed explanation on the model will be covered in Chapter 7.

## 2.10 IMPORTANCE OF THE SSD FOR REDUCING ENERGY, CARBON, AND WATER FOOTPRINTS

Strumiłło et al. [50] in *Handbook of Industrial Drying* (2014) discussed broadly the energy aspect of overall drying systems and methodologies to optimize the energy consumption for diverse type of dryers including SSD. As discussed in this chapter, energy reduction is needed to be seen holistically and reduction of energy is also related to improvement in carbon footprint. Further, in drying, evaporation of water is considered as loss of useful resource and requires to be studied in details for probable measures to recover it appropriately.

In this context, SSD is very much suitable in all three aspects: (i) the energy from the exhaust of steam can be possible to recirculate back into drying chamber (exhaust heat recovery), (ii) the energy absorbed during the drying mainly utilized for evaporation of water content in the feedstock is also possible to be recovered by use of excess steam suitably in the plant, and (iii) the water lost in the form of evaporated steam can be either used as heat transfer fluid or can be recovered after condensation recovering energy also. More discussion on this is given in Chapter 6.

It is likely to have higher heat transfer rate than HA and hence drying rate with the help of SSD [1,7,26,51,52] above inversion temperature ranging from 160°C to 300°C [3,28,53,54]. Nevertheless, all types of feedstock do not always prefer higher temperature of drying; certain heat-sensitive products need to dry at very low temperature (60°C–80°C). Various researchers experimentally showed that SSD improves the drying rate and efficiency as compared to mainly hot air dryers.

## 2.11 COMPARISON OF SSD AND HOT AIR DRYING

As SSD is a new drying media to replace conventionally used HAD, there are significant studies that discovered great benefits of it. The HA and SS properties are given in Table 2.2 for the comparative analysis with effect of a particular property on the drying phenomena from various perspectives. The important studies comparing SSD and HAD with their typical advantages in addition to limitations are discussed in the *Handbook of Industrial Drying* by Mujumdar [28] as well as comparative studies carried out in literature between SSD and HAD based on the various attributes such as drying kinetics mainly drying rate and time, energy utilized, product quality, else of processes may be due to oxygen-free environment or combination of multiple processes, etc. are represented in Table 2.3.

**TABLE 2.2**
**Properties of the HA and SS at 150°C**

| Items | Viscosity $10^{-6}$ kg/m s | Density kg/m³ | Specific Heat kJ/kg °C | Enthalpy kJ/kg | Heat Transfer Method | Heat Recovery Method | Fire Safety | Limitation |
|---|---|---|---|---|---|---|---|---|
| Hot air | 24.07 | 0.8 | 1.018 | 632 | Convection | Heat exchange | Oxygen dilution | Lower energy efficiency |
| SS at Atm | 14.18 | 0.52 | 1.98 | 2776 | Phase change (pre-condensation), convection, radiation | Condensation-latent heat | Oxygen free | Leakage at feedstock entry and exit |
| Inference on SSD | Pressure drop↓ velocity ↑ | Density ↓ dry spread ↑ | Thermal energy content ↑ | | Rapid-phase change; dominant: convection | Latent heat recovery ↑ | Completely Safe | Leakage prevention |

**TABLE 2.3**
**Comparison between SSD and HAD**

| S.N. | Paper Details with Authors | Type of Work/Dryer Type/ Feedstock | Parameters Compared SSD/HAD | Remark |
|---|---|---|---|---|
| 1. | Chu et al. [19] | Experimental and analytical Experimental test rig with sand | Drying rate | • Under same mass flow rate and temperature condition, higher drying rate for SSD but for same Reynolds Number and fully developed turbulent flow, drying rate is same |
| 2. | Yoshida and Hyodo [22] | Experimental wetted-wall column apparatus/water drops | Inversion temperature | • At higher temperature (above inversion temperature) the rate of water evaporation increases as the humidity of air increases and vice versa.<br>• Since water evaporates more rapidly into superheated vapor than into hot air above inversion temperatures, an ideal closed-circuit dryer should be operated at a temperature above it with SS for higher drying rate requirement. |
| 3. | Trommelen and Crosby [48] | Experimental study/falling droplet drying test apparatus/droplet of pure water and four food products and, five miscellaneous substances | Drying rate/product quality: thermal degradation | • At low temperature (150°C), hot air drying is significantly effective having higher drying rate but at high temperatures (250°C), both media are having nearly similar drying rate and no significant difference in product quality except particles dried by SS may be denser.<br>• In constant rate period, HAD is superior and fast than SSD, however for number of products in falling drying rate period, SSD is fast. |
| 4. | Bond et al. [39] | Experimental study of comparison between superheated steam and hot air impingement drying/ paper | Fluid jet of drying media: Temperature 150°C–450°C, GSM (g/m$^2$) 30–150 | • Compared to hot air drying, in superheated steam drying, the lower equilibrium moisture is achieved.<br>• Superheated steam impingement drying (SSID) is controlled mainly by internal transport resistance and not by adsorption of water on fibers especially in falling drying rate region.<br>• Critical moisture content for both drying medium rises with increasing the drying rate. It is less sensitive in SS than HA.<br>• Laboratory experimental results for SSID are encouraging and may be tried on pilot scale. |

(*Continued*)

**TABLE 2.3 (*Continued*)**
**Comparison between SSD and HAD**

| S.N. | Paper Details with Authors | Type of Work/Dryer Type/ Feedstock | Parameters Compared SSD/HAD | Remark |
|---|---|---|---|---|
| 5. | Moreira [60] | Mathematical modeling of impingement drying of tortilla and potato chips | Drying rate, nutritional value, temperature and medium velocity | • Impingement SSD can produce potato chips with less color deterioration and less nutritional losses (vitamin-C) than drying with hot air.<br>• Potato chips dry faster at high superheated steam temperature and have high convective heat transfer coefficients. |
| 6. | Wathanyoo et al. [61] | Experimental/fluidized bed test rig for paddy operating in batch | Head rice yield, whiteness, white belly, viscosity of rice flour and change of microstructure of rice kernel | • For the same drying time, drying rates of SSD paddy were lower than those HAD due to an initial steam condensation<br>• Promoted starch gelatinization improves head rice yield for SSD<br>• Whiteness of SSD paddy was lower than HAD around 2%. |
| 7. | Prachayawarakorn et al. [62] | Experimental and analytical study/fluidized beds test rig/ soybean | Drying rate, inactivation of antinutritional factors, protein solubility | • Comparing HAD operating in 135°C–150°C range with SSD operating below 135°C shows higher protein solubility of treated sample in latter.<br>• For the moist soybean, the types of heating medium do not impact protein solubility. |
| 8. | Stokie et al. [63] | Experimental and modeling study/fluidized bed drying test rig/Victorian brown coals: morwell, yallourn, and loy yang | Drying kinetics-drying rate and time, moisture readsorption/variable parameters: temperature-100°C, 130°C, 170°C for HA and 130°C, 170°C, 200°C for SS; velocity of drying medium-0.47, 0.57, 0.67 m/s for HA and 0.32, 0.48, 0.61 m/s for SS; particle size-0.5–1.7 mm | • Reduction in drying time by rise in temperature and velocity of drying medium, and decreasing particle size is observed for HAFBD and SSFBD indicating similar trend for drying kinetics.<br>• For same velocity, temperature, and particle size, drying time is almost same for HAFBD and SSFBD.<br>• Being steam oxygen free medium for drying, high temperature drying is possible and at this temperature (200°C), drying time lower for SSFBD than HAFBD (170°C).<br>• Between the HAFBD and SSFBD; the drying rates are proportion for similar variation in temperature and velocity of drying medium but not for change in particle size.<br>• Moisture readsorption is lower for SSD than HAD. |

*(Continued)*

**TABLE 2.3 (*Continued*)**
**Comparison between SSD and HAD**

| S.N. | Paper Details with Authors | Type of Work/Dryer Type/ Feedstock | Parameters Compared SSD/HAD | Remark |
|---|---|---|---|---|
| 9. | Hanifzadeh et al. [64] | Simulation-based study/ comparison of cow manure management technologies: SSD, HAD, etc. | Cow manure: indicators-energy payback time, life cycle assessment. configurations: SSD, HAD, field application, anaerobic digestion, eutrophication and global warming potential | • Thermal energy supplied by combustion of dried cow manure, helping reduction of fuel costs.<br>• Compared to field application, eutrophication and global warming potential is reduced by 95% and 70%, but, acidification potential is 35% higher for SSD.<br>• Higher environmental sustainability (70% lower impact) and 87% lower Energy payback time than anaerobic digestion for SSD indicating favorable future investment. |
| 10. | Brar et al. [5] | Experimental and optimization study/ experimental test setup/ yellow peas | Functional and nutritional properties: processing temperature, moisture content, hydration capacity, cooking characteristics, protein content, starch gelatinization, milling characteristics, and microstructural changes | • Lower reduction in protein content for SSD<br>• SSD minimizes overall cooking time of yellow peas without compromising functional and nutritional properties compared to HAD. |
| 11. | Jittanit and Angkaew [65] | Experimental-Laboratory-scale fluidized-bed dryer HAD and SSD; Two stage-first stage: SSD and second stage: HA oven drying and FBD; parboiled rice | Chalkiness and low milling, gelatinization of starch | • Hot water soaking combined with multi-stage intermittent drying method using SSD provided parboiled rice products with comparable or superior quality.<br>• Superheated-steam drying combines steaming and drying steps into one consumes less energy if the exhaust steam from SSD is recycled. |

### 2.11.1 Comparing Based on Drying Kinetics

Major differences between SSD and HAD are mentioned based on drying kinetics as follows: feedstock is supplied mainly at ambient condition, which is exposed to SS resulting pre-condensation in initial heat-up period causes high heat transfer represented by sudden rise in the temperature of the feedstock, which is not the case in HAD. Though feedstock in SSD above atmospheric pressure especially in the constant rate period is at saturation temperature of steam (sufficiently high temperature, 100°C) than wet bulb temperature for HAD (lower than SSD, approx. 60°C) the nature of the process is similar. Subsequently, transfer of surface moisture from the feedstock into the bulk steam can have negligibly small resistance being same medium. Additionally, the higher duration of the constant drying rate period is contributing toward lowering drying time as the greater drying rate will be present for longer duration in SSD than HAD. Subsequently, during falling drying rate period, moisture mass and vapor transport inside the feedstock have lower resistance for steam than HAD. Since the inversion temperature has a significant impact on the drying rate and time, it is discussed in the following subsection as demonstrated below.

#### 2.11.1.1 Effect of Inversion Temperature on Drying Kinetics

The inversion temperature is defined in several ways openly in literature. In the simplest form, it is the temperature where the rate of the drying of feedstock (dehydration of the feedstock) matches perfectly for both, SS and HA media, which is the widely recognized definition. Further, above inversion temperature, the drying rate is higher for SSD and it will increase with a rise in this temperature than HAD. However, for lower temperature than inversion temperature, it is vice versa. Because of this, it is preferred to have SS temperature above the inversion temperature benefiting higher drying rate in SSD than HAD. Costa and Da Silva [53] presented the fundamental study on inversion temperature for primary drying media, namely HA, humid air (HA and water vapor combination), and SS. It is reported that water evaporates quickly in SS than humid and dry air environment if media temperature is above the inversion temperature (170°C). This is described in Figure 2.11 for the above-mentioned drying media at atmospheric pressure in form of the graphs of drying rate vs. temperature of media like SS, hot moist, and hot dry air. At inversion temperature, drying rates for the above-mentioned drying media are equal and, as temperature increases further, the variation of drying rates are as follows: highest for SS then highly moist air, and lowest for dry air. Thus, inversion temperature is one of the important parameters detecting the drying rate for a given drying process and can be used for comparison between SSD and HAD.

Based on the understanding of inversion temperature, to have a higher drying rate and lower drying time for moist feedstock, for example solvents, ceramics, effluents, etc., temperature of SS at atmospheric pressure drying is ideally above 250°C higher than inversion temperature range for most of the substances. It is worth to mention that inversion temperature can be reduced using indirect heat supply, e.g., radiant heating mode can decrease the inversion temperature to 170°C from 250°C. Pronyk et al. [55] explained many potential benefits for the industries and consumers of use of SS as a drying medium for several products. Overall, SSD has significant energy savings potential (maximum 50%–80%) than HA or flue gases essentially because

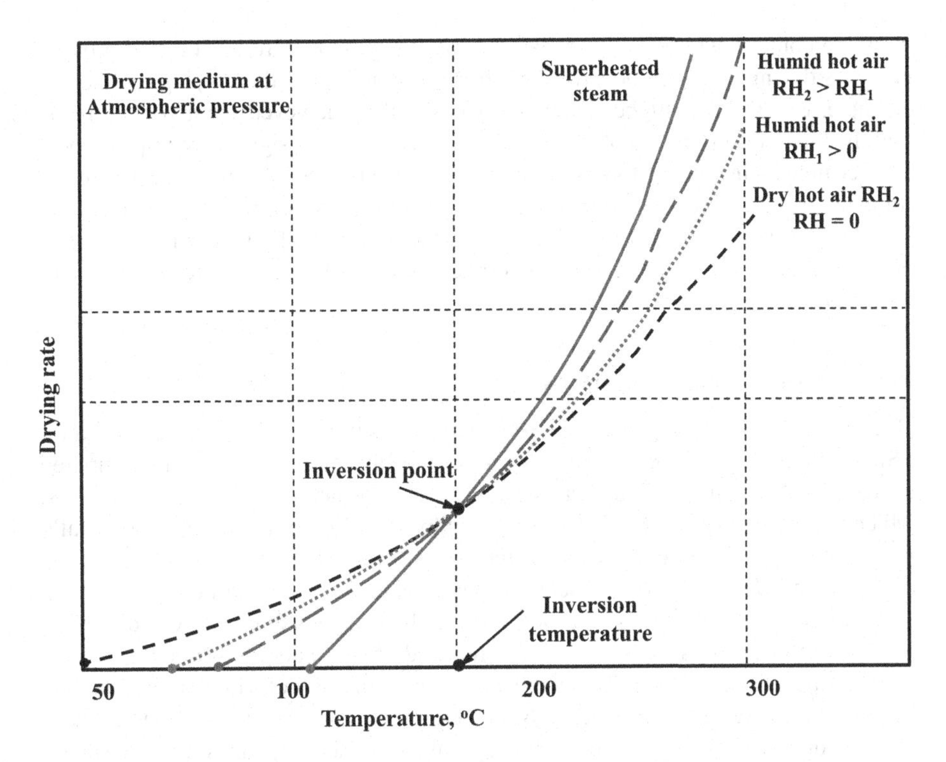


FIGURE 2.11 Variation of the drying rate for superheated steam and hot air with different humidity at atmospheric pressure corresponding to their drying temperatures. RH, relative humidity.

of greater heat transfer coefficients as well as drying rates in the constant and falling periods provided SS temperature is kept above the inversion temperature [55].

### 2.11.2 Energy Recovery and Resource Conservation by Recirculation in SSD

In closed operation of SSD, the thermal energy supplied is equal to the superheating of the steam recirculated back into the drying chamber and energy lost is only through the excess steam equal to the evaporation of moisture from feedstock, if it is not reused appropriately. Being closed system, it is easy to process the exhaust vapors, so separation of the impurities/pollutants by passing the surplus steam through the condensers [48] as well as recovery of the certain valuable volatile organic substances containing drying material and further segregation by a condenser is discussed by Tang and Cenkowski [56].

### 2.11.3 Product Quality and Oxygen-Free Surrounding

The control of quality parameters for the dried products is easily performed by the steam temperature regulation because they are dependent on the drying rates and final moisture content, which is also reliant on the steam temperature. The starch-based products drying using SS generally get better quality than other drying media. For example, due to SSD, higher head rice yield will be achieved compared to HAD mainly due to faster growth of starch gelatinization, causing significant improvement in intermolecular binding forces among starch granules [57]. The oxygen-free environment avoids oxidization, which may be encouraged by native enzymes resulting in colored compounds, deteriorating the product quality [54]. The SSD of biomass provides probability to soften the wood tissue, which is beneficial afterward for palletization [6,58].

### 2.11.4 Miscellaneous Differences

Overall and above, drying rates or lower energy consumption, the favored criteria for SSD are dried product quality, safe operation, and environmental concerns compared to that of HAD. Superheated steams are more appropriate for solid fuels like coals, which reduces the risk of fire hazards and explosion. Moreover, inert gases generally provided to switch on-and-off the system to safeguard from burning or fire while drying them. Some of the products are not valuable enough like hog fuel, sludge, peat, municipal solid wastes, etc. that may readily combustible in HA and need to be dried in huge mass (tons), where the reduced energy and environmental effects are most beneficial to use SSD since these improve green footprint due to reduction in pollutants ($CO_2$, and toxic gases $NO_x$, $SO_x$, etc.). But to get better cost competitiveness due to higher capital and very high mass of excess steam, it (excess steam) should have practically viable utilization appropriately in or nearby plant [57].

### 2.11.5 Droplet Drying

Trommelen and Crosby [48] performed series of experiments for drying of droplets (size 2 μL) of pure water and of a number of aqueous solutions and suspensions (four food products and five miscellaneous substances) with SS and HA. They have observed that, at low temperature (150°C), HAD is significantly effective having higher drying rate but at high temperatures (250°C), both medium are having nearly similar drying rate and no significant difference in product quality except particles dried by SS may be denser. Further, in constant rate period, HAD is superior and fast than SSD; however, for a number of products in falling drying rate period, SSD is fast. Another work proposed by Lum [59] on SS spray drying indicates better product quality for milk drying compared to HAD. Further, experiments with other materials representing proteins, carbohydrates, and common salts indicate that spray dryer with SS has significant potential to be used for unique features of dried products, which was not observed with HA media. In the case of common salts, superheated steam spray drying (SSSD) can be used for crystallization control forming hollow hopper-like sodium chloride crystals. Further, with mannitol finer and greater single crystals per particles are generated for SS compared to HA at same temperature conditions. Overall, as discussed before, SSD has multiple benefits and thus needs to be explored fully to get its benefit.

## 2.12 CLOSURE

In this chapter, initial discussion on physics of drying phenomena with SS media has mainly three regions viz., first, pre-condensation, where on one side there is condensation of steam on surface of the feedstock increasing the moisture to be evaporated by 8%–12% (may be beneficial in certain cases as favorable crystallization on food products); on other side, temperature of feedstock rises sharply due to higher heat transfer rate caused by condensation of the steam. This sharp rise in temperature may have sterilization effect on food products. Second, constant drying rate region which includes the evaporation of moisture condensed in the first pre-condensation region. In this case, product temperature remains constant equal to the saturation temperature of the steam at chamber pressure (e.g., for atmospheric pressure drying, it will be 100°C). Third, falling drying rate period, temperature of the feedstock increases and may be reaches the temperature of SS, but due to lower evaporation rate inside the feedstock, drying rate is reducing continuously. This region is especially important due to higher drying time and need special attention if drying time need to reduce. After this, advantages and disadvantages of SSD are discussed in more detail with applications for various types of substances. The chapter is closed with advantages of SSD over HAD. Overall, the SSD is having huge potential due to its unique benefits such as superior product quality, lower environmental concerns as being closed system, else of recovery from exhaust steam, and better control over exhaust steam utilization. Though there is considerable developed of SSD in the last two decades in lab-scale experimental work to find out new application and at pilot scale to make it commercially feasible further efforts are needed to tap its full potential. In the next chapter the application of SSD for biological products will be discussed with few case studies.

# 3 Superheated Steam Drying for Biological Products

## 3.1 INTRODUCTION OF SSD FOR BIOLOGICAL PRODUCTS

The biological products can be categorized as food products and non-food products. Primary sources of the food products naturally available are plants and animals, and all processed foods are derived from these only. They are generally classified based on life, processing to make edible, uses, applications, nutritional value, etc., as shown in Figure 3.1.

In general, all the food products used in day-to-day life, whether raw or processed, require the preservation to increase their shelf life, which slows down or stops spoilage. This will make food items available round the year with appropriate nutritional properties as well as use them in value-added products. The food decomposition or degradation may happen at the time of handling or may be because of physical, mechanical, microbial, or chemical damage, and among them, microbial and chemical decomposition are regular reasons [2,66]. Several enzymatic and chemical transformations during storage and processing of foods, e.g., browning, cause reduction in

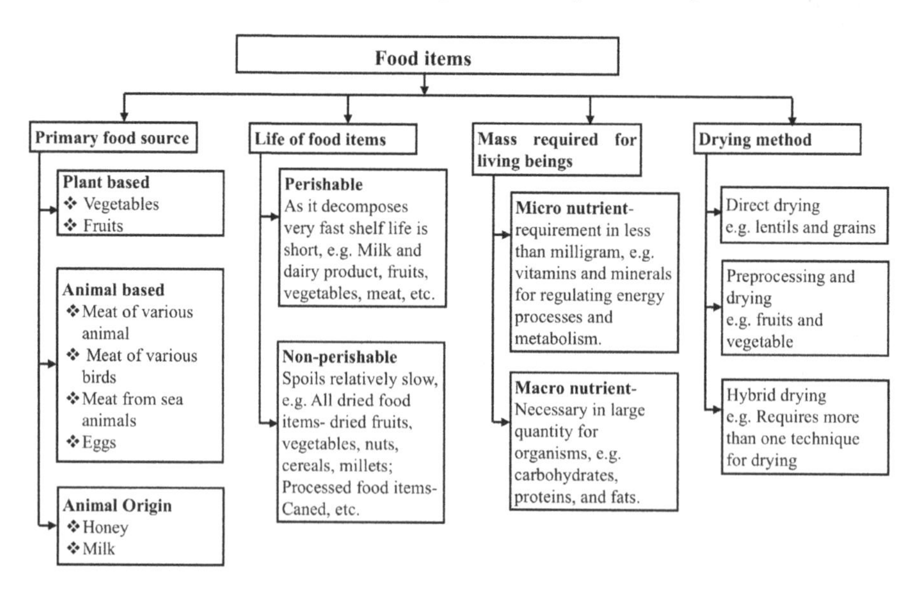


**FIGURE 3.1** General classification of the food items.


DOI: 10.1201/9781003275299-3

sensory properties and, in certain case, drop in nutritional qualities, resulting in inappropriate for consumption. Normally, food preservations are performed by various methods such as vacuum packing, freezing, canning, food irradiation, preserving in syrup, adding preservatives, and well-known thermal dewatering or drying. Drying of the food items improves significantly quality of them and also helps in diversifying recipes having delicious as well as nutritious food items. The biggest advantages of dried foods are that they can be easily stored in most of the cases without the need of special facility or container, used for value addition or to form different recipes easily, and may require lower storage space than products processed other than frozen foods or canned.

Drying is considered low-cost solutions available for preserving foods of all variety for removal of moisture present in feedstock by applying thermal energy. Several food items subcategorized are conserved using drying viz. meat products, marine products, and plants-based products. One of the latest techniques is superheated steam drying (SSD) due to its unique advantages especially for the food in addition to the already mentioned related economic benefits [67,68]. Several biological materials contents moisture even more than 90% in water form (e.g., maximum 93% moisture content in watermelon) which need to be decreased lower than critical level such that microbial growth can be restricted by appropriate technology. Limiting water content varies from microorganism to microorganism and is addressed in the literature based on water activity [66]. Further, drying of food products is subjective having separate approach based on various aspects of the features of feedstock's with appropriate pre- and post-processing step(s) so selection of most appropriate dryer is important for product quality [2,26]. The details of drying terminology, hygrothermal properties of vegetables, and fruits are given by Jangam et al. for ready references [69].

Plant-based food products contain grains, pulses, fruits, vegetables, spices, herbs, grasses, etc. where mainly fruits as whole or only seeds, leaves, flowers, roots and entire things including roots of these plants having applications as food products, medicinal values, or rich in mineral, vitamins, proteins. Because of these benefits, they need to be preserved or their utilization value may be increased by appropriate drying technology. Other animal-based food items are mainly various types of meat products like fish, pork, poultry, etc., which are also dried for increased life and other applications. Non-food biological products include biomass derived from agriculture residue and food processing industries, such as sugarcane bagasse, rice husk, beet pulp, empty fruit bunches, etc. In addition to drying, SSD also contributes to the pasteurization, disinfection, and sterilization of food products owing to oxygen-free or low-oxygen environment of steam along with a rapid rise in temperature above the safe temperature for microbes subjected to product feasibility. Further, it can be combined with either pre-processing or post-processing treatment such as blanching [62,70–72], parboiling [73], puffing or popping [74], etc. Usually, products dried in SSD are substantially porous than the hot air dried, since moisture is removed due to the boiling of water within the feedstock during SSD. This property is beneficial helping rapid reconstitution of dried product upon immersion in water. This is one of the vital characteristics of any "instant" food. Table 3.3 summarizes the important contributions from open literature for biological products. Overall, SSD has many advantages regarding the bioproducts to improve the product quality, utilization value, life, etc., which are discussed in more details in this chapter.

### 3.1.1 Pre-processing of the Biological Materials

The biological products available in the natural form have the effect of the surroundings such as dirt, dust, and other forms of foreign particles due to air, water, soil, insects, etc. Further, to enhance the drying properties and retention of the nutrients, it is important to have some sort of the natural or chemical treatment on them. Therefore, before drying, there is a need for pre-processing to be performed on the biological material. The following steps are followed in general before drying:

#### 3.1.1.1 Preparation of Biological Product Suitable for Drying

i. Discard rotten and otherwise spoiled materials.
ii. Remove dirt, dust, soil particles, etc. by soaking and washing with (hot/cold) water.
iii. Remove or peel-off the skin or outer cover, seed, and unwanted part, which is not necessary for drying depending on the type of biomaterials. For trimming, a flame treatment, hand knife, or abrasion peeler may be chosen suitably, for example, onion peels may be removed by charring and washing in a vibrating tumbler as a preferable or finer option.
iv. Chop, juice, or form puree as per the type of drying material.

#### 3.1.1.2 Pretreatments

i. Natural or bio-chemical treatment may be performed for better drying such as soaking salt or sugar water, etc.
ii. Pretreatments such as soaking, alkaline dip, blanching, freezing, sulphiting, heat treatment by thermal, high-pressure treatment, microwave treatment, and high-intensity electric field pulses (HELP) have been reported by several researchers. This enhances the dewatering characteristics, suppresses enzymatic reactions, and decreases the negative effects in the tissue of biological material during drying.
iii. Because of pretreatments, storage stability of dried products and the rate of drying increase, and also it may result in better textural, color, and organoleptic quality.

However, there are a few disadvantages such as protein denaturation, disruption of tissue cell membrane due to blanching, poor firmness and turgor loss, loss of color, flavor, and nutrients owing to heat treatments, etc. More details can be found in various textbooks on food processing [2,69,75,76].

### 3.1.2 Product Quality Attributes

The quality parameters of dried food products are categorized as: physical, biological, chemical, and nutritional [76]. The quality attributes generally considered to analyze the drying efficacy are tabulated in Figure 3.2.

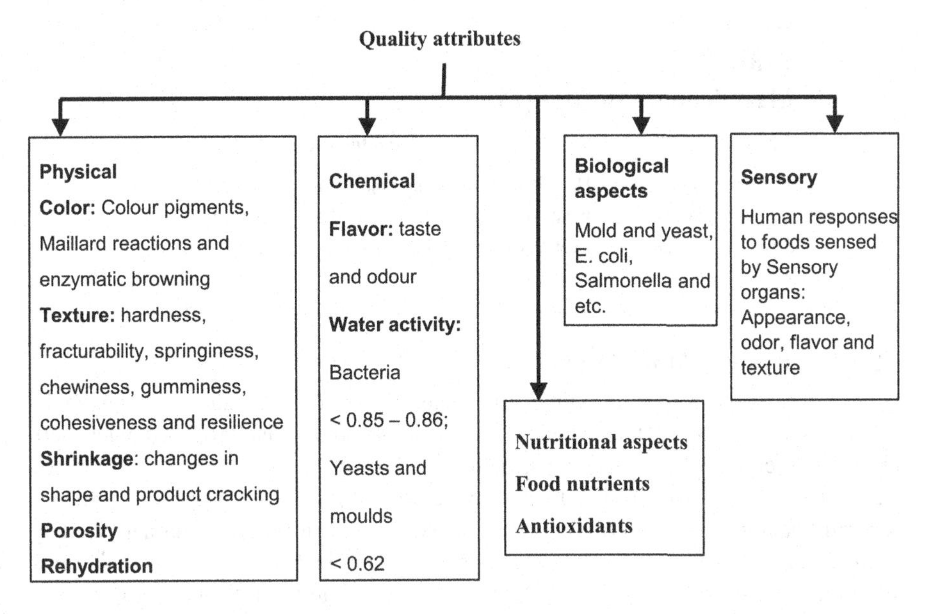


**FIGURE 3.2** Product quality attributes to analyze the drying process for biological products.

## 3.2 BIOLOGICAL PRODUCTS DRIED WITH SUPERHEATED STEAM (SS)

Researchers all over the world have tried drying diverse food products either at laboratory scale or pilot tests by using SSD to investigate the feasibility of SSD technology and product quality externally and internally. There is huge market potential for SSD with diversified product ranges which eventually be tapped for commercialization. The list of the researchers with their work from the area of the SSD of biological products namely foods are presented in tabular form in Table 3.3.

### 3.2.1 SSD of Fruits

The water content in fruits varies from very low to almost 100% though most of the fruits contain more than 80% water. The fruits with very high moisture content get spoiled, if not treated and handled properly. Further, the living and metabolizing tissues constituted in fruits are generally sensitive to temperature and fruits exposed to high temperature may be difficult to handle physically, reducing the life span of it, hence immediate processing is must.

For the reduction of water activity, various drying methods can be performed to get different forms of same fruits (or blended fruits) as slices, granules, cubes, and powders or leathers. For this different drying technology and equipment are required. Atmospheric pressure superheated steam-dehydrated products have better color and rehydration capacity compared to microwave vacuum-dried products [35]. Fruits are categorized based on their physical characteristics, viz. pome, stone berry, and tender, as shown in Table 3.2 [75], for which drying is explained in the following section.

**TABLE 3.1**
**Classifications of Fruit [75]**

| S.N. | Fruit | Characteristics |
|---|---|---|
| 1 | Stone | Hard inner layer around the seed |
| 2 | Pome | Fleshy receptacle tissue around a central core of seeds |
| 3 | Tender | Leathery rind and parchment-like partitions |
| 4 | Berry | Fleshy pericarp with seeds |

#### 3.2.1.1 Drying of Stone Fruits

These types of fruits contain pulp at the center and are further categorized as fruits with high rancidity, fruits that are sensitive to heat, and fruits that have hard outer shell. The high rancidity stone fruits may be better dried by SSD as it is oxygen-free environment and the heat-sensitive stone fruits may be dried at low-temperature SSD, which is generally operated at vacuum pressure [67]. Most of the stone fruits characterized by relatively harder shell are resulting in a higher drying time because it resists the transfer of moisture from feedstock as well as heat transfer into it. In that case, appropriate pretreatment can be chosen on a case-to-case basis. In this case also, SSD may be useful for combining the pretreatment with drying, improving the efficiency of the overall system and being energy-efficient technique, it may be economical [67] (Table 3.1).

Exotic stone fruits generally have the following characteristics:

- Hard outer shell
- High astringency
- High in vitamin C
- Rancidity
- Major polyphenol compounds are polyphenol oxidase (PPO) substrates

There are so many different types of dryers using conventional technologies that are currently in use; however, still there are certain challenges related to improvement in nutritional values, color, texture, drying time, etc. It may be possible that new appropriate drying technology may provide the required feature [1]. There is definitely certain potential for SSD technology based on the few studies listed in Table 3.3A, either by alone or by hybrid approach (combination with other drying methods).

**Case Study 3.1: Longan without Stone by Namsaguan and Mangmool [77]**

Longan is a seasonal fruit, economically important and commercialized in Thailand, China, Taiwan, and Vietnam though available in number of countries from Asia-Pacific region. It comprises high quantity of vitamins C and A, minerals such as phosphorus, iron, potassium, magnesium, as well as abundant in antioxidants. Initial moisture content is very high, 350%–500% d.b. The local and global availability of it is restricted due to its highly perishable nature (postharvest life 3–4 days), short shelf life, and vulnerability to postharvest diseases. For increase of storage life or value addition by drying it, traditional common practice followed by farmers or native businesspersons is hot air drying. However, this drying technique has challenges viz. long drying period, high energy usage, and the decline in product quality (like deterioration in color, taste, and nutritional constituents in numbers and quantity of product) [77,78]. Commercial hot air drying at 55°C–80°C of peeled and unpeeled longan required approximately 12–15 h and 48–72 h, respectively, to obtain 18% moisture content d.b. of dried product. Further, it is worth to note that only SSD at atmospheric pressure (temperature >120°C) would reach the product temperature 100°C resulting in burning of the longan even though drying time is significantly reduced for final moisture 18% d.b. of dried product. In this work, LPSSD is employed at lab scale to observe the effects of temperature and pressure of SS on drying kinetics and quality of dehydrated products indicated by various parameters such as shrinkage, color, rehydration, and texture (toughness).

###### *3.2.1.1.1 Effect of the Pressure and Temperature of SS on Drying Kinetics*

i. Drying rate: For a given pressure (lower than atmospheric), with increase in temperature from 70°C to 90°C, drying time is reduced significantly and, with reduction of pressure (15–7 kPa), drying is also reduced but for higher temperature of steam margin of drying time is lower than for low temperature. It means lower the pressure and higher the temperature, the minimum will be drying time. It is obvious that higher temperature means large temperature difference between SS and feedstock, providing a higher driving potential for drying. In case of lower absolute pressure, the boiling point of water is also lower increasing the driving potential for diffusion of the moisture outward. This can be interpreted as: the surface temperature of the longan is the same as boiling point, increasing the temperature difference between the SS and longan in a constant drying rate region will result in higher drying rate at lower pressure and higher temperature.
ii. Color of the dried longan: Comparing the influence of pressure and temperature of SS, the latter has greater effect on the color of the dried longan. Lightness does not affect due to temperature but, with increasing pressure, dried longan becomes darker. Overall, the change in color for 80°C and 90°C is more yellowish than 70°C (Table 3.2).

**TABLE 3.2**
**Consequence of Temperature and Pressure of SS on Drying Time of Longan**

| | Drying Time (min) | | |
|---|---|---|---|
| **Drying Pressure kPa (ab)** | **70°C** | **80°C** | **90°C** |
| 7 | 470 | 400 | 350 |
| 10 | 560 | 425 | 370 |
| 15 | 730 | 470 | 410 |

iii. Shrinkage and rehydration performance: With higher temperature, the shrinkage is lower and rehydration ratio is higher. This is due to higher temperature gives higher moisture removal resulting more porous structure and higher rehydration ratio. For 7 kPa (ab), shrinkage is lower compared to higher pressure (10 and 15 kPa) may be due to lower drying time and higher drying rate, resulting more porous structure. For higher pressure, the rehydration ratio is higher by significant margin. With higher pressure, porous structure may be collapsed, but number of the pores inside remains higher increasing the rehydration.
iv. Texture: Higher temperature and pressure of SS give tough longan after drying. For lower temperatures (70°C and 80°C) with all pressures, variation in toughness is not much. Similarly for lower pressures (7 and 10 kPa (ab)), variation is very small compared to 15 kPa (ab) in toughness.

Overall, lower pressure (7 kPa (ab)) and higher temperature (90°C) will have slightly lower drying time and shrinkage, but marginally higher color change, lower rehydration, and toughness.

---

### Case Study 3.2: From Husen [79] of Engkala Fruits and Avocados

The SSD can have high potential applications in underutilized fruits such as engkala fruits (padi variant) and avocados and compared with freeze drying (FD) process conventionally used. The results are discussed as follows. The performance parameters for investigations are used as the vitamins and mineral contents. Both fruits (engkala fruits and avocados) are dried at different temperatures (130°C, 150°C, and 170°C) in the superheated steam oven in laboratory. Drying time is very short for SSD (few hours) than FD (in days), saving significant amount of energy.

Engkala-has significantly higher total flavonoid content (TFC) (35% higher) and total phenolic content (TPC) (70% higher) particularly pulp dried by the SSD than the FD. This could be due to the application of heat changes in the chemical structure transformed the molecules to become polyphenols, which is fevorable. In contrast, the considerably lesser TPC (50% lower) and TFC (65% lower) noticed in the seed part dried in SSD than FD directly signify that the polyphenols, one of the constituents of seed, are mostly heat-sensitive, therefore may damage by higher temperature and energy of SS. For pulp, SSD gives significantly higher antioxidant capacity than FD. On the other side, FD seed of engkala fruits contains higher antioxidant capacity than SSD. Overall, SSD is appropriate drying technology for pulp of engkala fruits and for seed it is FD. Hybrid systems of SSD/LPSSD and FD/HAD may be complete solutions for entire fruit including the seed, which is interesting to try.

Avocados- pulp (edible part of avocado) gave significantly higher TPC and TFC after SSD, and its values rise with rising the superheat temperature. However, FD performed better with potentially greater content of TPC and TFC in the peels and seeds. It indicates that contents of TPC and TFC in the peels and seeds are highly affected by heat (temperature), thus loosed due to higher temperature and energy content in the superheated steam. Further, it clearly specifies mainly in the seed where TPC drops with the increase in steam temperature. For antioxidants, trend is similar to engkala fruits, SSD performing better than vacuum drying (VD) for pulp and other part seed and peel, VD is showing impressive performance.

Overall, in both fruits, engkala and avocados, the main edible part can be better dried by SSD at higher temperature (170°C) and remaining part of it can be processed by FD. It may be interesting to try only LPSSD or hybrid drying LPSSD with VD/FD to have complete drying solution.

---

#### 3.2.1.2 Drying of Pome Fruits

Pome fruits have a central core seed with fleshy surroundings (e.g., ber, acelora, snake fruit, etc.) or so many seeds like guava. The outer part of the flesh has a hard texture. They are classified based on their characteristics as: (i) heat-sensitive-recommended LPSSD, (ii) hard texture-may require pretreatment before drying, and (iii) thick outer skin- may be removed before drying.

#### 3.2.1.3 Exotic Tender Fruit Drying

It has normally seedy, soft, sticky flesh with a strong flavor. Generally, it has protein, fats, and compounds in considerable quantity but these are heat-sensitive.

---

**Case Study 3.3: From Jamradloedluk et al. [80] of Durian Chips**

Durian is a well-known and costly fruit, considered as king of the fruit mainly in tropical region of South-East Asia. These fruits are high in vitamin A, protein, carbohydrate, fat, iron, and phosphorous. Being a seasonal fruit, excessive supply of durian drops the prices significantly in peak time of production. Therefore, attempts are made for preservation and value addition. The atmospheric SSD is good option studied in details and compared with HAD. Figure 3.3 shows the experimental apparatus consists steam generator supplying saturated steam at 31.5 kg/h, which is superheated in an electric superheater (13.5 kW) to get superheated steam for drying in drying chamber. The exhaust steam is circulated by backward-curved blades centrifugal fan of capacity 2.2 kW motor. It also constitute of other supporting peripherals like pipes, valves, controls, measuring device, and data recording systems (not shown in Figure 3.3). Drying samples can be inserted by swing doors placed in the drying chamber. Test rig uses two drying media viz., hot air and superheated steam. Before passing the superheated steam, preheating of the complete system will be done by hot air till drying temperature to reduce condensation of the steam. Once drying temperature is reached, superheated steam is supplied appropriately as a drying media. The inlet temperatures of superheated steam vary as 130°C, 140°C, and 150°C and velocity 2 m/s during lab experimental trials.

Mathematical modeling for drying kinetics of durian chips:

Durian slice (after drying called chips) was considered as a 1-D slab in Cartesian coordinate. Fick's second law of diffusion (Equation 3.1) is used to curve fit the experimental drying kinetics data.

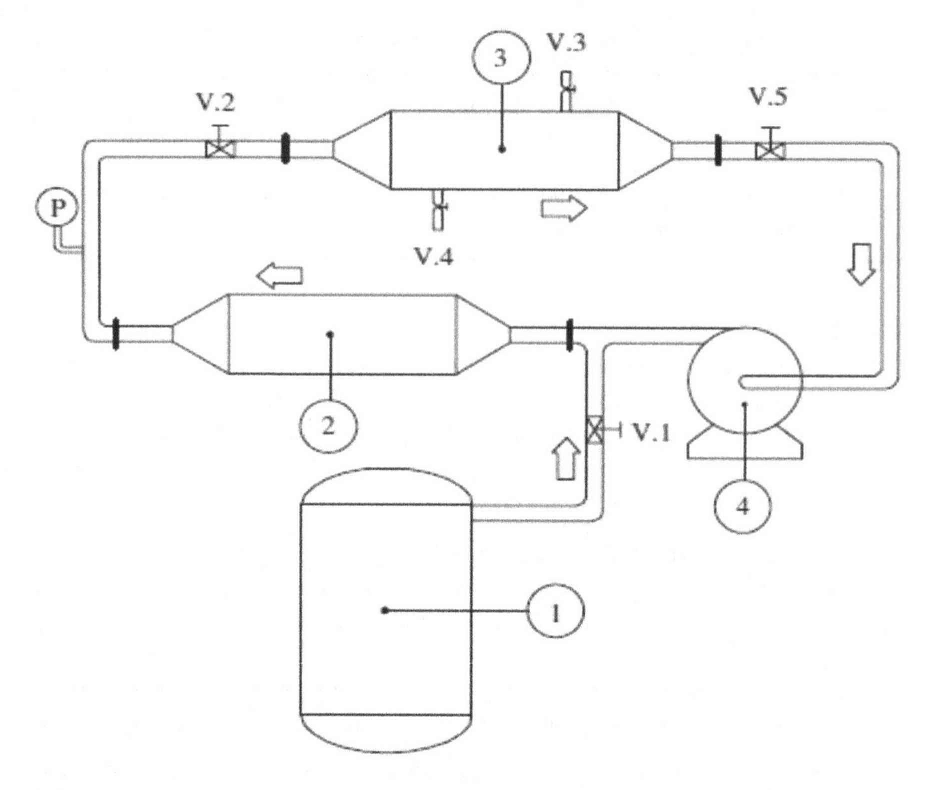


**FIGURE 3.3** A schematic diagram of the dryer: (1) boiler, (2) electric heater, (3) drying chamber, and (4) electric fan [80].

$$\frac{\partial M}{\partial t} = D_{\text{eff}} \frac{\partial^2 M}{\partial x^2} \tag{3.1}$$

where $x$ is the durian slice thickness (m), $M$ is the moisture content (d.b.), $D_{\text{eff}}$ is the effective diffusion coefficient ($m^2/s$) assumed to be constant, and $t$ is the drying time (s).

Solution of above equation given by Crank [81], consider infinite slab with spatial uniform moisture distribution and neglecting external mass transfer resistance is reproduced here as:

$$\text{MR} = \frac{M - M_e}{M_i - M_e} = \sum_{n=0}^{n=\infty} \frac{8}{(2n+1)^2 \pi^2} e^{\left(\frac{D_{\text{eff}(2n+1)^2 \pi^2 t}}{\left(L/2\right)^2}\right)} \tag{3.2}$$

where $M_i$ and $M_e$ are initial and equilibrium moisture content in slice respectively; maximum slice thickness $L$ is considered. The effective diffusion coefficient, as a function of the drying temperature, is determined by Arrhenius-type equation as:

$$D_{\text{eff}} = D_0 e^{\left(\frac{-E_a}{RT}\right)} \tag{3.3}$$

where $D_0$, Arrhenius factor; $R$, universal gas constant (8.314 J/mol K); $E_a$, activation energy in kJ/mol; $T$, temperature in $K$ are assumed.

The effects of using these two different drying media at different temperatures on the drying kinetics, color change (browning, yellowness, and lightness), compression, stiffness, hardness, rehydration capacity, and microstructure of the finished products were evaluated. The lower drying rates of SSD compared with HAD were observed due to pre-condensation during an early stage of drying as well as lower temperature than inversion temperature. On the adverse side, products dried by superheated steam dryers are having greater values of yellowness, redness, and rehydration capacity but lesser values of lightness.

From the SEM images it was observed that the SSD-dried product had less uniform, fewer but larger pores than hot air-dried products. The influences of the drying media and temperature on compression of the dried chips are not significant ($p > 0.05$). Overall, SSD at atmospheric temperature is giving better product quality but with change in color. Further, the effect of SSD on nutritional values and taste needs to be studied in detail and may be compared with HAD.

#### 3.2.1.3.1 *Remarks for Exotic Fruits*

To prevent oxidation, decomposition and change of flesh color and texture, conventional drying techniques are not suitable to dehydrate these fruits. It requires low or mild temperature drying techniques may be LPSSD without or with hybrid drying by FD, heat pump drying. In addition, it is interesting to investigate the change of volatile constituents of tender fruit in the future.

#### 3.2.1.3.2 *Remark on Fruit Drying through SSD*

There are good numbers of fruits where SSD/LPSSD is appropriate form of drying due to multiple reasons, however are not yet tested even at laboratory scale for drying such as dragon fruit (*Hylocereus undatus*), ciku/sapota (*Manilkara zapota*), Genipap (*Genipa americana* L.), saskatoon berries, Passion fruit (*Passiflora edulis* v. *flavicarpa*, *Amelanchier alnifolia*), papaya, citric fruits, watermelon, and so on. Looking toward the potential of the SSD, it is high time to go for sustainable drying for fruit preservation and processing.

### 3.2.2 SSD of Vegetables

It can be classified based on the use of the part of the plant (flowers, leaves, fruits, roots, or entire plant without/with roots) for cooking the vegetable dishes or can be taken raw as: (i) roots-based such as carrots, beets, garlic, onion, potato, sweet potato, parsley, ginger, etc.; (ii) leaves-based such as coriander, fenugreek, Korean traditional actinidia, cabbage, mustard greens, etc.; (iii) flowers-based such as cauliflower, broccoli, artichoke, banana flower, moringa, etc.; (iv) fruits-based such as yellow pea, bitter gourd, cucumber, pumpkin, tomato, peppers, etc.; and (v) entire plant-based such as palak, asparagus, lemon grass, etc.

Most of these vegetables are seasonable, available only 2–3 months in a year or, in unfavorable or hard climate, they are very costly. However, prices drop severely during peak season, making production uneconomical to farmers. Therefore, it is important to improve its shelf life by preservation as well as to do value addition for the expansion of its applications mainly by appropriate drying process. Few case studies are discussed in the following paragraph:

## Case Study 3.4: From Pimpaporn et al. [82] of Potato Chips

Generally, potato chips are prepared by deep-fat frying thin potato slices to moisture content of 0.02 kg/kg (d.b.) or less; however, the oil content of chips generally ranges from 35% to 45% (w.b.). Alternate to high level of oil content unhealthy snack products is organic or all natural, low-fat, low-calorie, low-sodium, low-carbohydrate have health-promoting benefits potato chips, which can be possible by SS drying. For conventional hot air drying of potato slices, critical issues are poor dissatisfaction texture, product color, and much nutritional degradation. SSD is recently applied for it, which is discussed in this section.

The potato chips being heat-sensitive need to be dried in LPSSD, which is offered superior choice being heat-sensitive foods and bioproducts. The experimental work is performed in test setup with pressure 7 kPa and temperature 70°C, 80°C, and 90°C. It is dried up to final moisture content of 3.5% (d.b.) from initial moisture content approx. 500% (d.b.). It is also important to note that pretreatment on the food material is playing major role in the drying along with final quality of product [66,82], which is not the scope of this book. The temperature of drying media is having considerable influence on the moisture removal rates from the samples. Higher the temperature of drying media, higher will be drying rate and lower drying time. As shown in Figure 3.4 for 90°C and 7 kPa pressure, drying will be 84 min to reach equilibrium moisture of 2% (d.b.). As the superheated steam at 90°C is passing over the potato chips kept at 0°C, there is pre-condensation of steam gaining the small amount of moisture, which is typically observed in SSD and LPSSD [26,70]. The large amount of thermal energy evolved due to pre-condensation of the steam is absorbed by sample (feedstock), suddenly increasing the

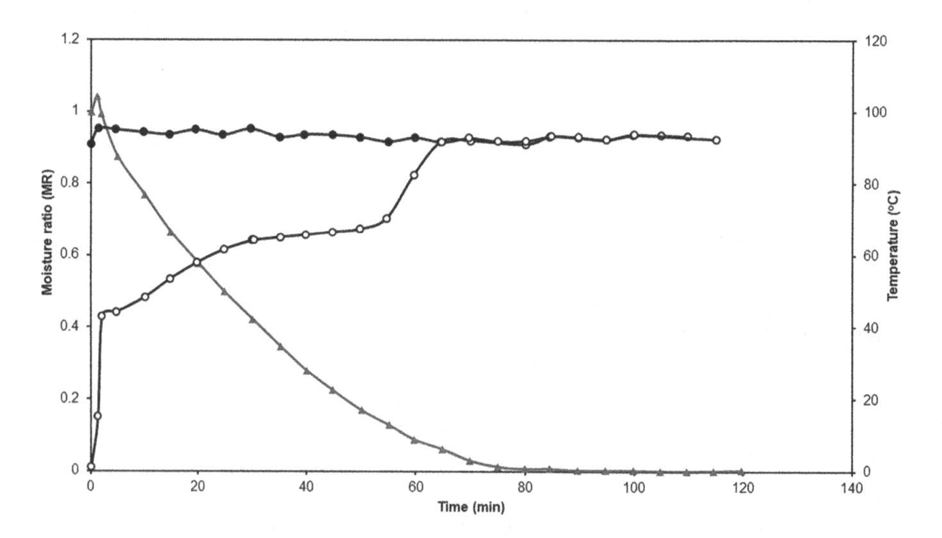


**FIGURE 3.4** Drying kinetics of dried potato chips going through LPSSD at 90°C (7 kPa): drying medium temperature (●), sample temperature (○), and moisture ratio (▲) [82].

temperature from initial to the saturation temperature of the water approx. 40°C (at 7 kPa pressure) and will remain at this temperature as constant drying rate region begins from this point. Afterward, completion of constant drying region, again temperature of the sample increased up to the superheated steam temperature. It is obvious that with increasing superheated steam temperature (70°C–90°C), there is significant rise in the drying rate and reduction in drying time (260–84 min). The reason behind is the large superheat degree leads to higher driving potential for transfer of heat/mass. Overall, in the case of SSD, greater the superheat temperature, higher will be moisture diffusion and hence moisture removal rate.

---

#### 3.2.2.1 Effect on the Potato Chips Quality

i. **Color**: With the change in the superheat temperature, the changes in lightness, redness, and yellowness are investigated appropriately. The drying temperature of the superheated steam has not showed any substantial influence on the changes of lightness in different types of pretreatment as mainly gelatinization of starch takes place during the pretreatment itself. Overall, temperature does not influence the redness of dried samples, however, based on the pretreatment, for few configuration, higher redness is observed at particularly 90°C superheat temperature than other lower temperatures (70°C and 80°C) due to the possibility of Millard reaction or other thermal damage. As compared to fresh and pretreated samples of potato chips, the yellowness is increased for SS-dried samples, may be because of rise in carotenoid content per unit weight during drying. The drying temperature variation does not have any substantial effect on the yellowness irrespective of pretreatment.

ii. **Effect of superheat temperature on texture**: The effects of superheat temperature on textural parameters of dried potato chips such as final thickness, hardness, crispness, and toughness are also studied. For change in thickness of dried samples, mainly temperature is not having any significant effect, but based on treatment, at 90°C temperature, higher thickness is observed than other temperatures. Lower temperature 70°C is having higher hardness compared to other higher temperatures (80°C and 90°C). At higher temperatures, probably puffing improves the porosity, reduces hardness, and less shrinkage of the samples (representing the thickness of the chips after drying). Similarly, the dried potato chips are having lower value toughness and higher crispness with increasing temperatures from 70°C to 90°C.

In conclusion, blanching and freezing together, without any chemical pretreatment followed by drying in a LPSSD at a temperature of 90°C with 7 kPa absolute pressure was recommended as the best conditions for producing potato chips in this study. At this parameters, better product quality, shorter total drying time and less thermal damage than other conditions are reported. On the other side, a sensory study of dried potato chips is must, before arriving at any decision on the validity of the results presented here.

### Case Study 3.5: From Brar et al. [5] on Yellow Pea

*Pisum sativum* L. called popularly yellow peas are rich in proteins, so superb replacement for animal-based protein, high content of several nutritions like complex carbohydrates, soluble and insoluble fiber, vitamin B, and minerals such as iron, calcium, and potassium. It is used in diverse food items such as in preparation of soups, roasted to get crispy nut-like snack, and to improve dietary and functional properties of food items, pea purees is best mix. Other use, processed pea and isolated proteins are also used in baking mixes, baked goods, breakfast cereals, soup mixes, health foods, processed meats, and pastas. The consistent high-quality product of peas with appropriate color, flavor, and texture is possible by particular quality of it. Currently for yellow peas, the average cooking time is 15–20 min if soaked at room conditions for 24 h. The market demand of yellow peas is based on the cooking quality of it especially which is used in soup cuisines.

The SSD is used for the drying of the yellow peas at 120°C, 135°C, and 150°C processing temperatures with velocity 1 m/s. The important properties of yellow peas like moisture content, hydration capacity, tempering, cooking characteristics, starch gelatinization, characteristics (dehulling), protein content, milling, and microstructural changes are studied appropriately to find the optimal process of drying of it. The SSD is designed and developed by previous researcher used in this experimentation shown in Figure 3.5 [83]. Overall, investigations are conducted to develop an efficient processing with same nutritional qualities of the yellow peas for minimal cooking time.

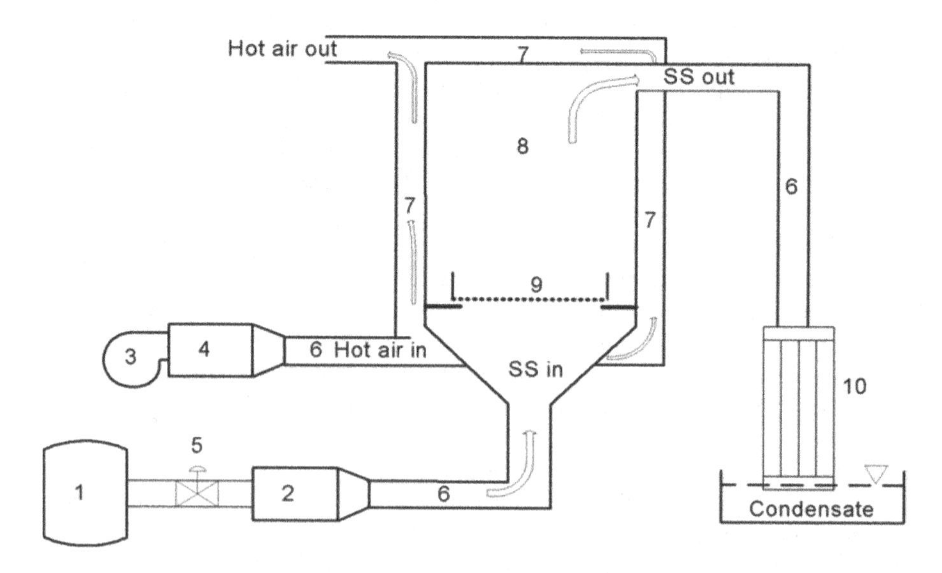


**FIGURE 3.5** Line sketch of the superheated steam drying (SSD) systems [83]. (1) Steam generator; (2) superheater; (3) fan; (4) air heating chamber; (5) steam flow control valve; (6) SS or hot air conveying pipes; (7) hot air jacket; (8) SS processing chamber; (9) sample tray; and (10) condenser.

i. **Drying characteristics:**

Similar to Iyota et al. [36], the drying time reduces with increasing the temperature of superheated steam temperature is observed in case of yellow peas. Also, for higher moisture peas (@ 54%) required little more time (15 min) to achieve equilibrium moisture i.e.12% than the peas of lower initial moisture content 26% (10 min) at same SS temperature. Overall, for all temperatures (120°C–150°C), the effect of initial moisture content in peas on drying time is uniform and relatively small (an average 10 min). The maximum pre-condensation (@ 3% rise in moisture w.b.) is noticed for peas dehydrated at 120°C, however practically no initial condensation is noted with superheated steam, 150°C and 1 m/s velocity.

ii. **Cooking characteristics**:

Initial moisture in the peas has substantial effect on the textural parameters while cooking of the peas. For peas of 26% initial moisture content dried with SS for all temperatures (120°C, 135°C, and 150°C) cannot be cooked appropriately even after 10 min. For SSD peas with higher initial moisture content, cooking time reduces with increasing drying temperature of SS and have better quality of products. The optimal cooking conditions (lowest cooking time, 5 min without soaking) are possible for yellow peas dried at 150°C SS and 54% (w.b.) initial moisture content.

iii. **Dehulling efficiency**:

It is inversely proportional to the equilibrium moisture content and directly proportional to processing temperature of the peas. The yellow peas processed at 150°C with SS have maximum dehulling efficiency (89.9%) and no non-dehulled fraction.

Protein content in yellow peas after drying with SS at all temperatures is reduced by very small margin due to non-oxygen media. The absence of oxygen, denaturation reactivity is almost negligible resulting in a very less reduction in protein content. There is no significant effect of temperature of SS observed on starch gelatinization and energy required for it during peas drying.

Microstructure analysis: The middle lamella breaking in SS-dried peas for all temperatures is appropriately observed in Figure 3.5. This breakage was observed due to a combination of steaming and sudden rise in the temperature leads to evaporation of high rate of water vapors causing a deformation of the middle lamella, losing of integrity of the cell. With rise in the SS temperature, the damage to the middle lamella of peas also rises ensuing in a crack between the cotyledons. Overall, all happening in the microstructure mainly due to high heat and moisture transfer within peas would similarly changes hardness.

The SSD peas at 135°C and 150°C have hydration capacity in the range of 80.7%–91.2%, which means more water absorption in the cotyledon compared to lower temperature dried. Increase in processing temperature also raise the rehydration capacity of yellow peas. This is happening especially for SS-dried peas as it causes favorable changes in microstructural increasing hydration capacity (as discussed in the above section) (Figure 3.6).

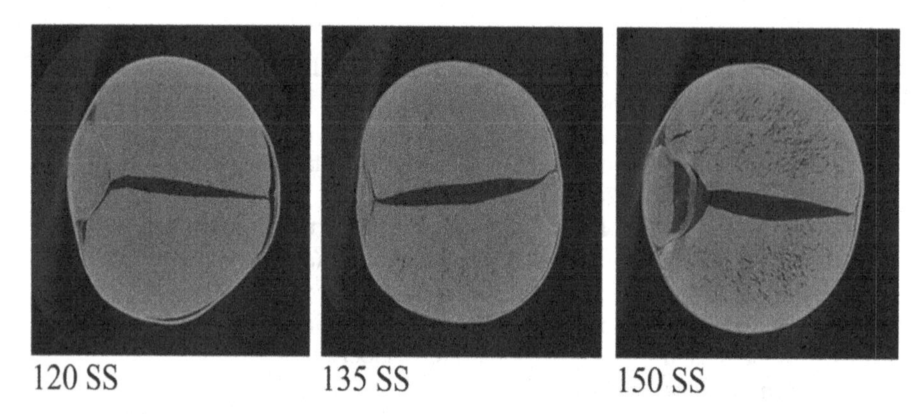


**FIGURE 3.6** 2-D images of single kernel of pea dried in superheated steam at 120°C, 135°C, and 150°C [5].

Further, in comparison with hot air dryers, it is observed that SS has a noticeably higher moisture removal rate, dehulling efficiency, better microstructure and porosity, and less nutrition loss, indicating a better choice for yellow peas. The optimal condition for SSD of yellow peas is a superheated steam temperature 150°C, 1 m/s steam rate and 54% initial moisture (w.b.) in peas for lower cooking time (5 min) and higher drying rate or lower drying time (15 min).

### 3.2.3 SSD of Roots

Root is part of a plant body mainly present underground, having the main purpose of storage of nutritive food. The various types of roots based on their shape and pattern are tuberous, taproots, corm, tuber, bulbs, and rhizome. The concentration and the balance between sugars, starches, and other types of carbohydrates are different in different roots. However, they are perishable and have the adverse effect of high temperature, causing them to droop giving a poor appearance. Thus, the preservation of roots is critical for keeping them long time without much deterioration in the quality. Among various approaches, dehydration by drying is widely used to improve the shelf life of roots. Before drying begins, the pre-processing, mainly cleaning and shaping in addition to pretreatments, are essential for the roots. In this section mainly applications of SSD for roots are discussed.

---

**Case Study 3.6: From Suvarnakuta et al. [84] on Carrot**

Carrot is one of the best sources of the β-carotene, which is antioxidant and anticancer, and also a precursor of vitamin "A." However, it is seasonal though yield is taken in ample amount and also, especially in instant food industry, dried carrot is preferred for various dishes. LPSSD is proposed to dry carrot with minimal loss of the nutrients such as β-carotene, and vitamin "A." The experimental investigations on lab scale test rig developed by Suvarnakuta et al. [84] on Carrot are proposed in the following paragraphs.

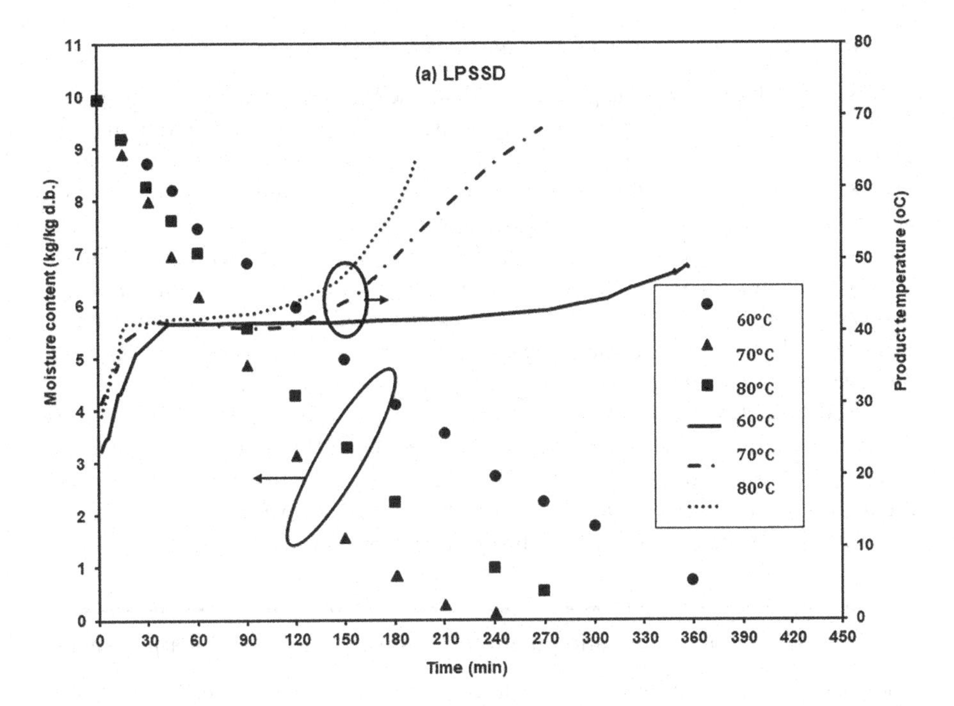


**FIGURE 3.7** Drying kinetics of carrot for various SS temperatures at 7 kPa (ab) pressure [85].

Drying kinetics: The experiment is performed for 7 kPa (ab) pressure, 60°C, 70°C, 80°C temperatures, and 26 kg/h velocity of the superheated steam. The carrot is stored at 4°C and dried from initial moisture content 9.5–10 kg/kg (d.b.) to final moisture content 0.1 kg/kg (d.b.). As temperature of the superheated steam increases (60°C, 70°C, 80°C) the drying time significantly reduces (420, 330, 210 min) as shown in Figure 3.7 [85]; however the quality of the dried product also reduced.

**β-carotene retention**: It is observed that with lower temperature, the retention of the β-carotene increases, and at 60°C, it is highest (@ 0.83), however drying time is quiet high (420 min). The β-carotene in carrot is highly sensitive to oxygen content in the drying media and temperature, as the SS is oxygen-free, the β-carotene is higher and at low temperature it is highest. The carotene is degraded by free radical oxidation mechanism and it depends on the heating temperature and time, and quantity of oxygen. Further, the activity of lipoxygenase and peroxidase causes the oxidative degradation of β-carotene and shows its effect at higher drying temperatures resulting in lower β-carotene at 70°C and 80°C. For the case of 60°C SS, higher product temperature, close to 60°C, occurs after drying time, 300 min and moisture content, 2 kg/kg (d.b.), where reduction in β-carotene is observed due to thermal degradation.

Overall, LPSSD is better option compared than vacuum and hot air drying for carrot due to improved physical properties, i.e., better rehydrated and redder dried carrot for optimal operating conditions as 60°C temperature and 7 kPa operating pressure with 420 min drying time, even though it is comparatively higher than other vacuum and hot air drying.

### 3.2.4 SSD of Spices, Herbs, and Medicinal Plants

Spices and herbs are aromatic products with medicinal value, extracted from specific types shrubs to get taste to foods of all types made from either plant or animal origin in addition of health benefits [86]. There are hundreds of diverse types of spices all over world having special ingredient due to local environmental conditions including soil. Asian region is the prominent grower of various herbs and spices, predominantly coriander (cilantro), pepper, cinnamon, cloves, nutmeg, and ginger, etc.; in Europe mostly bay leaves, basil, chives, celery leaves, coriander (cilantro), thyme, dill tips, and watercress are grown. In America spices produced are pepper, ginger, nutmeg, allspice, and sesame seed. These are consumed fresh as well as dry, but former is more versatile in use due to else of handling and flexibility of mixing. However, there is a need to be taken care of steam drying of spices, herbs, and medicinal plants, which are having content of terpenes, it is having high affinity with steam and may lost with exhaust steam. Furthermore, number of spices, herbs and medicinal plants are required to be dried for increasing shelf life as well as utilization. In this section how best LPSSD can be utilized for the drying applications is discussed.

**Case Study 3.7: From Kozanoglu et al. [87] on Coriander and Pepper Seed**

The vacuum- or low-pressure superheated steam fluidized bed dryer is operated at lower pressure than atmospheric pressure. In this case study, superheated steam at pressure 40 kPa (ab) performs the fluidization of the feedstock to be dried [87]. The supplementary heat provided by the electric resistance heaters fitted on the external walls of the fluidized bed column helps to have the temperature of feedstock higher than the saturation temperature of steam at chamber pressure and hence the initial moisture gain is not observed for coriander and pepper seeds for all drying temperatures (exception 90°C for pepper seeds), however, there was reduction in the moisture from the beginning of the drying process for coriander seeds only for all temperatures and for pepper seeds except 90°C condition. Further, higher temperature of SS corresponding to higher degree of superheating, $\Delta T = T_o - T_{sat}$, helping to accomplish a drying process without pre-condensation (exception 90°C for pepper seeds). Additionally, effect of the variation of pressure on the drying rate and time shows with decreasing the pressure, drying time reduces mainly due to increasing the degree of superheat. For coriander the degree of superheat 2°C–4°C will have significant effect and with increasing it, no effect on final moisture contents.

## 3.3 CLOSURE

SSD is one of the important approaches for preservation and value addition of the food. In this chapter, the applicability of the SSD is discussed for drying of various food items, however lot of work is to be done to explore its full potential. The major advantages of the SSD other than commonly mentioned are being oxygen-free environment; it helps to reduce oxidation and other reactions, improving product quality in terms of higher nutrients, color retention, porous and better microstructure, better water reabsorption, natural sterilization, etc. Further, in certain food products, pre-condensation of steam improves the product quality acting as pre-processing. Moreover, for the heat-sensitive food items, LPSSD is a prospective option, discussed with its successful implementation through few case studies. However, industrial setup required for SSD is costly and complex due to supporting equipment like, steam boiler and systems for feeding and extraction of the feedstock and drying product, respectively. Overall, superheated steam is a better option for drying of biological products and its full potential needs to be explored (Table 3.3).

**TABLE 3.3**
**Summary of the SSD Applications for Biological Products**

**A. Fruits**

| S.N. | Paper Details with Authors | Type of Work/Dryer Type | Product/Quality Parameters | Remark |
|---|---|---|---|---|
| 1. | Shaharuddin et al. [88] | Experimental Lab scale with kinetic study. Superheated steam treatment at 170°C for 15 min prior | Baccaurea pubera/antioxidant properties and mineral content as well as to determine nutritional values | • SS treatment enhanced compared to the treatment, the total phenolic content by 147.8% and DPPH radical scavenging activity by 23.7%.<br>• Reductions, as compared to the control treatments, in total flavonoid content by 16.5%, lycopene content by 28.6% and ferric reducing antioxidant power by 22.2%.<br>• Reduce the mineral content of the fruit, from as little as 3.6% to as high as 52%. |
| 2. | Malaikritsanachalee et al. [89] | Experimental Lab scale/ analytical LPSSD continuous and intermittent | Ripe mango/color, shrinkage, rehydration, and microstructure | • Intermittent LPSSD was able to reduce the drying time, color changes, shrinkage, and rehydration time of dried ripe mangoes when compared with HAD, and has superior porous structure and high rehydration rate for dried samples. |
| 3. | Lim et al. [78] | Experimental lab scale/ with drying kinetics model verification | Asam gelugor/drying time, acid number | • The optimal conditions for superheated steam drying of Asam gelugor (G. cambogia) fruit rinds were identified as 46.6 min and 150°C with the composite desirability of 0.913.<br>• Application of superheated steam drying under controlled conditions resulted in faster drying process and better quality of dried G. cambogia than conventional sun drying technique. |

*(Continued)*

**TABLE 3.3 (*Continued*)**
**Summary of the SSD Applications for Biological Products**

**A. Fruits**

| S.N. | Paper Details with Authors | Type of Work/Dryer Type | Product/Quality Parameters | Remark |
|---|---|---|---|---|
| 4. | Namsaguan and Mangmool [77] | Experimental lab scale/ LPSSD: drying temperatures 70°C–90°C and absolute pressures 7–15 kPa. | Longan without stone/drying from 350% to 400% dry basis until 18% dry basis. Color, shrinkage, rehydration and toughness | • Drying rate increased with increasing drying temperature and with decreasing pressure.<br>• Drying time and product quality, LPSSD at 80°C and 15 kPa was the best.<br>• Total color change seemed to increase with the increase of drying temperature and with the decrease of pressure.<br>• Higher temperature provided better quality in terms of shrinkage and rehydration while lower pressure gave better quality in terms of shrinkage and texture. |
| 5. | Eang and Tippayawong [90] | Experimental lab scale/ parametric optimization-temperature, velocity and drying time | Cashew apple (testa) with nuts/ color change and process energy consumption | • For maximum yield of kernel and other quality indices, optimum thermal processing condition of superheated steam drying at 30 min of drying duration, 4 m/s velocity and 115°C temperature |
| 6. | Alfy et al. [73] | Review SSD for food/ low-pressure, fluidized bed heating, Hybrid/ combination with infrared radiation | Products reviewed-Zousoon, chicken meat, fish press cake; banana slices, mangosteen rind, longan, Indian gooseberry, carrot, potato; paneer, spent grains, paddy, soybean, tortilla chips; coriander and pepper seeds, basil leaves, chitosan film enriched with gooseberry extract and galangal extract | • Drying of food products, enzyme inactivation, decontamination, microbial load reduction, and parboiling are also essential and important for dryer selection.<br>• Features: product related factors: sensitivity to temperature and oxygen, moisture content, thermal resistance and taste/ aroma, process related factors: other uses of steam, environmental emissions from dryers, combustion/explosion hazards and cost of thermal energy.<br>• LPSSD is having vast potential for diverse applications due to preferred quality of products either higher valuable components retention. |

*(Continued)*

**TABLE 3.3 (*Continued*)**
**Summary of the SSD Applications for Biological Products**

**A. Fruits**

| S.N. | Paper Details with Authors | Type of Work/Dryer Type | Product/Quality Parameters | Remark |
|---|---|---|---|---|
| 7. | Husen et al. [79,91] | Experimental Lab scale/ comparison with freeze drying | Avocado/Temperatures (130°C, 150°C, and 170°C), total phenolic, total flavonoid, and antioxidant activity | • Improved total phenolic, total flavonoid, and antioxidant activity than freeze drying |
| 8. | Jamradloedluk et al. [80] | Experimental Lab scale/ comparison with hot air drying | Durian chips/drying kinetics-temperatures (130°C, 140°C, and 150°C), quality attributes, as well as microstructure | • Lower drying rates of SSD compared with HAD due to steam condensation occurred during an early stage of drying as well as the thicker dense layer formed in SSD.<br>• SSD provided products with greater values of redness, yellowness, as well as rehydration capacity but lower values of lightness than HAD.<br>• Effects of the drying medium and temperature on compression characteristics of the dried chips, however, were not significant. |
| 9. | Methakhup et al. [92] | Experimental lab scale/ comparison: LPSSD, VD | Indian gooseberry flake/drying kinetics, shrinkage, color, and ascorbic acid retention, temperature: 125°C, 150°C and 170°C | • Vacuum drying shorter time (nearly half 145 min at 75°C and 7 kPa (ab)) to dry the product than that required by LPSSD at tested drying conditions (temperature: 65°C and 75°C; pressure (ab): 7, 10, 13 kPa)<br>• Except for vacuum drying at 75°C and 7 kPa (ab), LPSSD could retain ascorbic acid better than the vacuum drying. In addition, LPSSD could preserve the color of the sample better than the vacuum drying at tested conditions. |

(*Continued*)

**TABLE 3.3 (*Continued*)**
**Summary of the SSD Applications for Biological Products**

**B. Vegetables**

| S.N. | Paper Details with Authors | Type of Work/Dryer Type | Product/Quality Parameters | Remark |
|---|---|---|---|---|
| 1. | Brar et al. [5] | Experimental/lab scale | Yellow pea kernels/moisture content, hydration capacity, cooking characteristics, dehulling efficiency, microstructure, starch gelatinization, and protein content | • SS dried Yellow pea gives superior quality in most of the parameters than hot air dried and further reduces the cooking time and prior to cooking soaking requirement is eliminated.<br>• The proposed work can be extended for commercial application. |
| 2. | Alp and Bulantekin [68] | Review | Microbial inactivation and microbiologically safe drying process | • Based on microorganisms deactivation, LPSSD is one of the most effective to decrease Salmonella spp.<br>• The effect of this is due to the steam temperature being over 100°C. |
| 3. | Chan et al. [93] | Experimental lab scale oven: comparison: SSD, commercially dried (CD), fresh available. | Labiatae herbs: rosemary, peppermint, thyme, oregano, sage, spearmint and marjoram. Temperature in oven 150°C and 200°C for 5, 10 and 20 min. | • SSD of rosemary, thyme and marjoram showed enhanced anti-tyrosinase properties for all the drying regimes, further in fresh marjoram, tyrosinase not detected in fresh samples.<br>• SSD may be a promising drying technique of Labiatae herbs for the commercial production of tyrosinase inhibitors. |
| 4. | Sehrawat and Nema [94] | Experimental lab scale mode, mathematical modeling for moisture prediction/multi-purpose drying unit: HAD, VD and LPSSD: Comparison | Onion slice/comparison: LPSSD, VD and HAD three temperatures (60°C, 70°C, and 80°C), total phenol content (TPC), antioxidant, pungency, color, and rehydration | • Low-pressure superheated steam drying at 70°C has been found the best drying condition.<br>• LPSSD able to retain high pungency, better color, and rehydration of dried onion. |

(*Continued*)

**TABLE 3.3 (*Continued*)**
**Summary of the SSD Applications for Biological Products**

**B. Vegetables**

| S.N. | Paper Details with Authors | Type of Work/Dryer Type | Product/Quality Parameters | Remark |
|---|---|---|---|---|
| 5. | Kim et al. [95] | Experimental/SSD with oven drying | Korean traditional actinidia (*Actinidia arguta*) leaves/total phenolics, flavonoids content, and antioxidant activity. | • Total phenolics, flavonoids content, and antioxidant activity increases with increasing drying time up to 160 s dramatically decreased at drying of 200 s.<br>• Aerobic bacteria were not detected at drying time over 120 s and coliform of all the samples was not detected.<br>• The superheated steam was very effective drying method of increase to the nutritional and sanitary quality. |
| 6. | Liu et al. [96] | Experimental lab scale mode/vacuum drying (9.5 kPa) and LPSSD (9.5 kPa) at 75°C–90°C (5°C interval) | White radish discs/drying kinetics and attributes-vitamin C, rehydration ability and microstructure | • As expected, the drying rate increased with an increase of drying temperature and at lower temperature drying time is lower for LPSSD than VD. But at higher temperature (90°C), it is nearly same for both.<br>• Due to loose structure and pores in the surface (porous microstructure), the good rehydration ability.<br>• Compared to VD, better color retention than VD. Vitamin C degradation is lower and also recovery from the exhaust superheated steam is also possible. |
| 7. | Phungamngoen et al. [97] | Experimental lab scale mode/HAD, vacuum drying (10 kPa) and LPSSD (10 kPa) at 60°C with pretreatment methods | Cabbage/effects of pretreatment and drying methods on the resistance of Salmonella, physical properties-color and shrinkage | • Drying without pretreatment could not completely eliminate Salmonella, while no salmonella was detected on the pretreated samples at the end of the drying process.<br>• No effect of pretreatment and drying methods on volumetric shrinkage.<br>• Dried blanched samples exhibited greener and darker color than the dried acetic acid pretreated and untreated samples |

(*Continued*)

**TABLE 3.3 (*Continued*)**
**Summary of the SSD Applications for Biological Products**

**B. Vegetables**

| S.N. | Paper Details with Authors | Type of Work/Dryer Type | Product/Quality Parameters | Remark |
|---|---|---|---|---|
| 8. | Kingcam et al. [98] | Experimental lab scale mode/LPSSD with pretreatment methods | Potato chips/degree of starch retrogradation, initial slice thickness (1.5, 2.5 and 3.5 mm), final moisture content, hardness, toughness and crispness, degree of crystallinity | • Higher degrees of starch retrogradation led to an increase in the degree of crystallinity, hardness and toughness of dried chips, but did not show any significant effect on the crispness.<br>• Higher initial thickness leads increase of hardness and toughness (reduction of crispiness) of potato chip. |
| 9. | Pimpaporn et al. [82] | Experimental lab scale mode/LPSSD with various pretreatments | Potato chips/temperatures (70°C, 80°C, and 90°C) at an absolute pressure of 7 kPa, colors, texture (hardness, toughness and crispness) and microstructure | • LPSSD at 90°C with combined blanching and freezing pretreatments is most optimal process<br>• Redness and yellowness of dried potato chips were not significantly affected by drying temperature but by the pretreatment methods.<br>• Freezing improves lightness and crispness and also reduce the toughness. |
| 10. | Leeratanarak et al. [72] | Experimental Lab scale/ comparison: LPSSD and HAD | Potato chips/blanching time, drying methods, drying temperature, color retention, texture (hardness) and degree of browning | • A blanching time of 5 min followed by LPSSD at 90°C at an absolute pressure of 7 kPa was proposed as the best condition.<br>• Blanching and drying temperature significantly affected the hardness of potato chips under certain conditions while the drying method did not show any significant influence on the hardness.<br>• The best condition proposed still lower hardness compared with the commercially available potato chips. |

(*Continued*)

**TABLE 3.3 (*Continued*)**
**Summary of the SSD Applications for Biological Products**

**B. Vegetables**

| S.N. | Paper Details with Authors | Type of Work/Dryer Type | Product/Quality Parameters | Remark |
|---|---|---|---|---|
| 11. | Suvarnakuta et al. [84] | Experimental lab scale/ comparison: LPSSD, VD, and HAD | Carrot/degradation kinetics of β-carotene | • Degradation kinetics of β-carotene depends more on temperature than moisture content in carrot in all types of dryers.<br>• LPSSD led to less degradation of β-carotene of carrot than hot air drying (up to 20%–25% higher in LPSSD).<br>• At lower temperature (60°C) but higher drying time, β-carotene is highest in LPSSD. |
| 12. | Pronyk et al. [55] | Experimental lab scale/ comparison: SSD, HAD | Potato, sugar-beet pulp, Asian noodles, and spent grains/ drying characteristics, drying rates, and the effect of superheated steam on product quality in thin-layers | • Sugar-beet pulp dried under the same drying temperature and velocity had the same water activity irrespective of drying medium, i.e. same drying medium temperature and velocity, the sorption properties do not depend on the drying medium used.<br>• During SSD the center temperature showed a sharp rise to 100°C during the warm-up period and then a gradual rise. A constant-rate drying period in superheated steam dried samples at 125°C and 145°C but no discernable constant-rate period at 165°C due to immediate surface hardening or increased thermal driving force |

(*Continued*)

**TABLE 3.3 (*Continued*)**
**Summary of the SSD Applications for Biological Products**

**B. Vegetables**

| S.N. | Paper Details with Authors | Type of Work/Dryer Type | Product/Quality Parameters | Remark |
|---|---|---|---|---|
| 13. | Caixeta et al. [99] | Experimental lab scale Impingement drying and analytical heat transfer coefficient/comparison: SSD and HAD, commercial, and fried potato chips | Potato chips/Parametric analysis (temperature 115°C–145°C; 100 and 160 $W/m^{2}°C$), Shrinkage, density, porosity, color, texture, and nutrition loss | • Potato chips produced using superheated steam impingement drying showed more shrinkage, higher bulk density, lower porosity, and lighter color than chips dried with air under the same temperature and with the same convective heat transfer coefficient (130°C and 145°C. $h=100$ $W/m^{2}°C$).<br>• Superheated steam dried potato chips retained more vitamin C during the drying process. |
| 14. | Iyota et al. [70] | Simulation with experimental lab test rig for initial moisture consideration | Potato slice/shrinking, swelling, moisture content and temperature distributions in a material, changes in mass of a material with time, and a characteristic drying curve, influence of the initial thickness of a material and the heat transfer coefficient | • Modeling of the condensation and evaporation occurring during the initial stage of superheated steam drying with regard to mass transfer within the material and its volumetric change.<br>• Using models calculation, the respective time changes for the temperature distributions, moisture content distributions, and mass of the material, characteristic drying curves were predicted to investigate the influence of condensation and validation with potato chips experimental results.<br>• the larger the combined heat transfer coefficient and heat flux given to the surface the shorter the restoration time and the smaller the maximum amount of condensate. |

(*Continued*)

**TABLE 3.3 (*Continued*)**
**Summary of the SSD Applications for Biological Products**

**B. Vegetables**

| S.N. | Paper Details with Authors | Type of Work/Dryer Type | Product/Quality Parameters | Remark |
|---|---|---|---|---|
| 15. | Moreira [60] | Experimental lab scale Impingement drying and analytical Heat transfer coefficient/SSD and HAD comparison | Tortilla and potato chips/color, nutritional losses (vitamin C), convective heat transfer coefficient | • Impingement drying with SSD can produce potato chips with less color deterioration and less nutritional losses (vitamin C) than HAD.<br>• Potato chips dry fast at high temperature of SSD (130°C) and has high convective heat transfer coefficient. |
| 16. | Van Deventer and Heijmans [100] | Experimental pilot scale SSD with heat recovery | Cauliflower and carrot/ feasibility, drying time | • Drying of two products was completed successfully in pilot scale SSD. |
| 17. | Tang and Cenkowski [56] | Experimental Lab scale/ parametric analysis and SSD and HAD comparison | Cylindrical potato/dehydration characteristics, temperature histories, drying rates, and overall moisture diffusivities of | • The temperature of SSD had a greater effect on all parameters.<br>• A constant-rate drying period was noticeable only in the dehydration with the SS of 125°C and 145°C, and was not present in the dehydration with the SS of 165°C and with the HA of 125°C–165°C.<br>• There existed an inversion temperature point between 145°C and 165°C for the first dehydration stage above 2.6 kg/kg db and between 125°C and 145°C for the last dehydration stage below 2.6 kg/kg db. |
| 18. | Li et al. [101] | Experimental superheated steam impingement drying lab scale/SSD and HAD comparison | Tortilla chips/temperature profile, drying curves, physical properties (shrinkage, crispiness, starch gelatinization and microstructure) | • The steam temperature had a greater effect on the drying curve than the heat transfer coefficient within the range of study however high temperature (>145°C in this case) shows high effect of heat transfer coefficient on microstructure and shrinkage.<br>• The samples dried at a higher steam temperature and a higher heat transfer coefficient had less fully gelatinized starch.<br>• At high temperature (145°C) drying rate is higher in SSD than HAD. |

(*Continued*)

**TABLE 3.3 (*Continued*)**
**Summary of the SSD Applications for Biological Products**

**B. Vegetables**

| S.N. | Paper Details with Authors | Type of Work/Dryer Type | Product/Quality Parameters | Remark |
|---|---|---|---|---|
| 19. | Bernardo et al. [102] | Experiment pilot plant/ comparison: hot air and SSD | Sugar-beet fiber/drying rate and dried product quality: Color, water retention capacity, SS velocity 0.1–2 m/s and temperature 120°C–150°C. | • Water retention capacity is constant and independent of the drying media.<br>• SSD temperature 130°C–150°C yields 90% of the dry matter but drying time is higher due to initial condensation.<br>• HAD 40°C–105°C does not change the original color of the beet fibers up to 90 of DM.<br>• The drying time 2.5–6 min HA and 10–15 min, SS with a flow rate of 1.7 m/s. SS might also be efficient as a deodoration agent. |
| 20. | Yoshida and Hyodo [47] | Experimental lab/SSD and HAD comparison | Potato chips/Drying rate, energy consumption, appearance and color | • SSD at 240°C with 14 kg/h, shows better appearance and color than air-dried under the same conditions<br>• SSD potatoes had 40% less oxidation than those fried with hot air.<br>• Potato chips are more porous and permeable to vapor than when dried in air drying. |
| 21. | Karrer [8] | Commercial oven (700 mm of Hg)/electric LPSSD oven | Cabbage/drying time, color | • Natural cabbage Color was maintained at 130°C SS temperature and 700 mm of Hg pressure time 5 h |

(*Continued*)

**TABLE 3.3 (*Continued*)**
**Summary of the SSD Applications for Biological Products**

**C. Other Food Products**

| S.N. | Paper Details with Authors | Type of Work/Dryer Type | Product/Quality Parameters | Remark |
|---|---|---|---|---|
| 1. | Ma et al. [103] | Experimental on model scale/comparison: superheated steam spray dryer and hot air spray dryer | Instant coffee and sodium copper chlorophyllin (natural green colorant) powders/ SSSD parameters: Mean diameter, bulk density, and solubility time<br>Process parameters: inlet temperatures (160°C–200°C) and feed rates (3–15 mL/min)<br>Product parameters: Yield, physical properties and morphology | • SSSD produces the natural green colorant powder with smaller particle size, higher bulk density, and more wrinkle surfaces, resulting in superior solubility.<br>• Instant coffee adhered and dried on wall chamber and is not suitable in the current situation<br>• Better for natural green colorant with improved properties.<br>• The yield and color, both of the powders and of the reconstituted solutions, no significant differences due to drying media either SS or HA. |
| 2. | Lee et al. [104] | Experimental/Analysis | Perilla oil Seed: Steam pretreatment before pressing/ Yield, viscosity, color parameters, acid value (AV), and peroxide value (PV), fatty acid composition and volatile profiles, TPC and antioxidant activity | • The SS treatment resulted in the damaged cellular structure of perilla seeds and furnished about 2.5-times more oil yield as compared to the yield without treatment.<br>• This is obtained without any adverse effects in the viscosity, color parameters, AV, PV, fatty acid composition, and volatile profiles.<br>• SHS treatment resulted in increased TPC and antioxidant activity about 5-times and a dramatic reduction of lipase activity of Perilla Oil. |

(*Continued*)

**TABLE 3.3 (*Continued*)**
**Summary of the SSD Applications for Biological Products**

**C. Other Food Products**

| S.N. | Paper Details with Authors | Type of Work/Dryer Type | Product/Quality Parameters | Remark |
|---|---|---|---|---|
| 3. | Choicharoen et al. [105] | Experimental test rig/ coaxial SS two-Impinging stream dryer (ISD) at atmospheric pressure (120 kPa ab) comparison: SSD and HAD | Okara (soy residue)/ performance of the drying system: volumetric heat transfer coefficient, volumetric water evaporation rate, temperature (130°C, 150°C, 170°C and 190°C) and velocity (20 and 27 m/s) of SS, material feed flow rate and dryer geometric parameters: impinging distance (5, 9, and 13 cm) | • The volumetric heat transfer coefficient insignificant effect on the inlet steam temperature, while increases with an increase in the inlet steam velocity, material feed flow rate, and impinging distance.<br>• An increase in the inlet steam temperature, steam velocity and material feed flow rate led to an increase in the volumetric water evaporation rate.<br>• Lowest total specific energy consumption of the system was around 3.1 MJ/kg water at an inlet steam temperature of 190°C, inlet steam velocity of 20 m/s, material feed flow rate of 20 kgdry solid/h, the impinging distance of 5 cm, and steam recycle ratio of 63%.<br>• For SSD, up to 46% and 95% savings in the total specific energy consumption compared with the cases of hot air with and without exhaust air recycle, respectively. |
| 4. | Gómez et al. [106] | Computational and experimental pilot scale study/Impingement jet drying | Rapeseeds/simulation with experimental results of moisture kinematics | • Simulation of a radial impingement jet dryer with SS in continuous operation shows good agreement between the Eulerian model based on kinetic theory of granular flow (KTGF) and preliminary experimental results. |

(*Continued*)

**TABLE 3.3 (*Continued*)**
**Summary of the SSD Applications for Biological Products**

**C. Other Food Products**

| S.N. | Paper Details with Authors | Type of Work/Dryer Type | Product/Quality Parameters | Remark |
|---|---|---|---|---|
| 5. | Prachayawarakorn et al. [62] | Experimental, lab, and analytical/fluidized beds test rig comparison: SSD, HAD | Soybean/drying rate, inactivation of antinutritional factors, protein solubility | • For temperature of 135°C–150°C for the HAD and below 135°C for the SSD, the SSD type heating medium shows the protein solubility of treated sample to be higher than HA to the dry soybean.<br>• For the moist soybean, the types of heating medium do not impact on the protein solubility. |
| 6. | Kozanoglu et al. [87] | Experimental Lab scale LPVFBSSD/pressure variation 40, 53, 66 kPa; Temperature of SS: 90°C, 100°C, 110°C. | Coriander seed and pepper seed/operating pressure and temperature | • The degree of superheating (2°C and 4°C) was identified as the most important parameter over the process.<br>• Higher drying rates and lower final moisture contents by increasing operating temperature (relatively lower between 90°C and 110°C).<br>• Moisture gain in the warm-up period of the process was prevented providing some supplementary heat to the fluidized bed through electrical resistances attached to the walls of the column. |
| 7. | Iyota et al. [74]; Yotaro et al. [107] | Experimental lab scale and analytical modeling/ fluidized bed | Amaranth seed popping/ Expansion ratio by popping and the type of heating media, gas temperature, initial moisture content of the seeds and heating time, texture | • Compared with the case of hot air, the expansion ratio by superheated steam was slightly lower.<br>• The relationship between the popping conditions and seed's temperature, moisture content and internal pressure changes can be determined by model.<br>• After popping, 8.7 times higher than the raw samples' volume of the amaranth. |

*(Continued)*

**TABLE 3.3 (*Continued*)**
**Summary of the SSD Applications for Biological Products**

**D. Rice/Parboiling/Paddy**

| S.N. | Paper Details with Authors | Type of Work/Dryer Type | Product/Quality Parameters | Remark |
|---|---|---|---|---|
| 1. | Junka et al. [108] | Experimental/analytical lab/batch fluidized bed dryer | Jasmine brown rice (JBR)/three drying media, SS, moist air, dry air/rancid odor elimination | • Thiobarbituric acid (TBA) concentration of the sample dried using SSD Fluid bed 150°C was the lowest over a 180-day storage period.<br>• However, high temperatures tended to harden cooked JBR owing to starch gelatinization.<br>• Moreover, it caused the 2-ace- tyl-1-pyrroline (2AP) volatile of the dried samples to decrease under all drying conditions. |
| 2. | Jittanit and Angkaew [65] | Experimental laboratory-scale/ multistage intermittent-hot water soaking combined-first stage: SSD and second stage: HA: oven at 50°C drying or FBD at 80°C.<br>Comparison: Two stage with single stage and conventional technique | Parboiled rice/chalkiness and low milling, gelatinization of starch | • The head rice yields improved from below the national standard (6.2% for Plai Ngahm Prachin Buri and 29.5% for Prachin Buri 2) to high milling quality levels (up to 61.4% and 68.2%, respectively) by hot water soaking together with multistage intermittent drying technique with high head.<br>• The superheated-steam drying combines the steaming and drying steps into one process and consumes less energy if the exhaust steam from SSD is recycled. |

(*Continued*)

**TABLE 3.3 (*Continued*)**
**Summary of the SSD Applications for Biological Products**

**D. Rice/Parboiling/Paddy**

| S.N. | Paper Details with Authors | Type of Work/Dryer Type | Product/Quality Parameters | Remark |
|---|---|---|---|---|
| 3. | Hampel et al. [109] | Modeling of single rice grain with experimental validation on lab scale: Transient continuum-scale Macroscopic using heat and mass transfer based on volume-averaging approach | Single brown rice grain/ absolute permeability and the bound-water diffusivity and drying kinetics, thermal conductivity and specific heat capacity, Particle density and porosity | • The native color of brown rice particles can be preserved if the drying temperature remains below 160°C during the superheated steam drying process.<br>• From a practical point of view, this model is rather complex and not readily suited for technical applications. |
| 4. | Chen and Xu [110] | Experimental Lab scale SSFB roasting device, heating and puffing, numerical simulation analysis of roasting process, Comparison: HAD/SSD | Chinese rice wine: one kind of non-waxy and two types of waxy/microstructure, pasting properties and starch crystal, optimize the operating parameters and enlarge the fluidized bed | • By numerical simulation, it is found that temperature gradient in rice is relatively low and temperature changes in a small range in the late during rice roasted at 200°C they have high gelatinization and lower surface browning rice, and be allowed to have broad residence time distribution in the fluidized bed. In comparison, rice roasted at 220°C, the temperature rises faster, but larger temperature difference in rice and temperature of drying medium is too high in the late, resulting browning, therefore, need to control the residence time in the fluidized bed.<br>• Comparison between air and superheated steam as two different media shows that the gelatinization effect of rice is practically the same given the same fluidized state, although superheated steam exhibits a slight advantage. |

(*Continued*)

**TABLE 3.3 (*Continued*)**
**Summary of the SSD Applications for Biological Products**

**D. Rice/Parboiling/Paddy**

| S.N. | Paper Details with Authors | Type of Work/Dryer Type | Product/Quality Parameters | Remark |
|---|---|---|---|---|
| 5. | Kozanoglu et al. [111] | Experimental/Fluidized bed test rig, parametric analysis by varying temperature (98°C–118°C), pressure (40–67 kPa ab), velocity (2.9–4 m/s) of SS | Paddy/drying kinetics-drying rate | • Increasing operating temperature enhanced the drying rates and decreased the equilibrium moisture content. Reducing the operating pressure had an influence to a lesser extent; however, it increased the degree of superheating, controlling the phenomenon, associated with the operating temperature as well as the operating pressure.<br>• Variation of the superficial steam velocity had a minimal effect over the transient moisture profiles during the constant drying rate period, while in the falling drying rate period almost no effect was observed.<br>• Increasing the degree of superheating enriches the drying rate in the constant drying rate period and produces lower equilibrium moisture contents. The degree of superheating higher than 30°C had nearly no effect over the process and should not be applied in the case of heat-sensitive materials.<br>• It is feasible to accomplish LP-SSFBD for paddy drying. |

*(Continued)*

TABLE 3.3 (*Continued*)
Summary of the SSD Applications for Biological Products

D. Rice/Parboiling/Paddy

| S.N. | Paper Details with Authors | Type of Work/Dryer Type | Product/Quality Parameters | Remark |
|---|---|---|---|---|
| 6. | Swasdisevi et al. [145] | Experimental model test rig/impinging stream dryer<br>Comparison: SS and HA medium for ISD<br>Inlet air temperatures: 130°C, 150°C, and 170°C; SS: 150°C, 170°C, and 190°C | Paddy/volumetric water evaporation rate, volumetric heat transfer coefficient, specific energy consumption, dried paddy: color, head rice yield, and starch gelatinization | • The SEC decreased with an increase in the drying temperature. The SEC for SS was lower than HA as the drying medium at the same drying temperature.<br>• Superheated-steam drying with 90% recycle could conserve more energy than with 60% recycle.<br>• The color of the dried paddy was not affected by the change in the drying temperature in all cases; superheated-steam drying led to slightly redder and more yellow dried paddy.<br>• The percentage of head rice yield decreased with an increase in the drying temperature; superheated-steam drying led to higher head rice yield than hot air drying when considered at the same drying temperature.<br>• The degree of starch gelatinization was highest for drying at 150°C for both SS and HA. |

(*Continued*)

**TABLE 3.3 (*Continued*)**
**Summary of the SSD Applications for Biological Products**

**D. Rice/Parboiling/Paddy**

| S.N. | Paper Details with Authors | Type of Work/Dryer Type | Product/Quality Parameters | Remark |
|---|---|---|---|---|
| 7. | Soponronnarit et al. [112] | Testing of a pilot scale (100 kg/h parboiled rice; 160 kg/h saturated steam) and mathematical model development/ superheated-steam fluidized bed dryer | Parboiled rice/changes in temperature of steam and moisture content, water adsorption, whiteness and pasting viscosities, white belly, and hardness | • Superficial velocity of steam from 1.3 to 1.5 times of the minimum fluidization velocity had no significant effect on the drying rates of rice.<br>• Energy consumption for reducing the moisture content from 0.43–0.22 kg/kg dry basis was approximately 7.2 MJ/kg water evaporated for 4–5 min drying time.<br>• Soaking paddy at a temperature of 70°C for 7–8 h before drying was sufficiently enough for producing parboiled rice, with no white belly.<br>• The gelatinization of starch during drying resulted in higher head rice yield than raw paddy by higher than 60 % for higher than 0.18 kg/kg dry basis moisture content. |
| 8. | Wathanyoo et al. [61] | Experimental/fluidized bed test rig for batch operation paddy, comparison with HAD | Paddy/head rice yield, whiteness, white belly, viscosity of rice flour and change of microstructure of rice kernel | • For the same drying time, drying rates of SSD paddy were lower than those HAD due to an initial steam condensation.<br>• Promoted starch gelatinization improves head rice yield for SSD paddy.<br>• Whiteness of SSD paddy was lower than HAD around 2 %. |

*(Continued)*

**TABLE 3.3 (*Continued*)**
**Summary of the SSD Applications for Biological Products**

**E. Processed Food Products**

| S.N. | Paper Details with Authors | Type of Work/Dryer Type | Product/Quality Parameters | Remarks |
|---|---|---|---|---|
| 1. | Srivastav and Kumbhar [113] | Expt/analytical (ANN) LPSSD | Paneer/prediction of moisture content, drying rate and moisture ratio | • To develop artificial neural network (ANN) models for drying kinetics of paneer using superheated steam drying methods. |
| 2. | Mayachiew and Devahastin [114] | Experimental lab test rig/ comparison: SSD drying temperatures (70°C, 80°C, and 90°C) at 10 kPa, Moist HAD, ambient air, vacuum drying | Edible chitosan films/drying kinetics, color, tensile strength, percent elongation, water vapor permeability (WVP), glass transition temperature (Tg), crystallinity, storage condition (%RH) on WVP | • Vacuum drying and LPSSD required much shorter drying time than did ambient and hot air drying at 40°C. However, LPSSD took a slightly longer time than vacuum drying.<br>• LPSSD at 70°C was proposed as the most favorable conditions for drying chitosan films due to less yellow color, higher tensile strength, and, percent elongation than the conventional HAD and vacuum drying. |
| 3. | Pronky et al. [115,116] | Experimental/ mathematical model of drying comparison: SSD, HAD, fried commercially available | Instant Asian noodles/kinetics, temperature, velocity, drying time, textural properties: Adhesiveness, springiness, cohesiveness, chewiness, resilience, and hardness | • Textural properties are unaffected by steam velocity<br>• Pretreatment with saturated steam increased hardness and chewiness.<br>• Increase in SS temperature decreased adhesiveness, springiness, cohesiveness, and resilience but slightly increased hardness and chewiness.<br>• Noodles processed at a steam velocity of 1.5 m/s and at 125°C for 200 s, 130°C for 167 s, 135°C for 150 s, 140°C for 133 s, 150°C for 100 s, and a steam velocity of 1 m/s and 150°C for 133 s had acceptable color values and moisture at or below the safe storage limit. Textural properties are comparable with commercially available. |

(*Continued*)

TABLE 3.3 (*Continued*)
**Summary of the SSD Applications for Biological Products**

**F. Hybrid Dryers: SSD with Some Other**

| S.N. | Paper Details with Authors | Type of Work/Dryer Type | Product/Quality Parameters | Remark |
|---|---|---|---|---|
| 1. | Romdhana and Bonazzi [117] | Experimental lab scale/ hybrid: SSD and HAD | Apple cube/color, drying time | • Proposed hybrid system for drying of apple cube to get good color and reduce drying time.<br>• SSD at 170°C reduced almost 50% of the initial moisture within 8–10 min otherwise it takes almost 80–180 min for hot air at 75°C–105°C |
| 2. | Somjai et al. [118] | Experimental lab scale/ mathematical modeling based on mass and energy balance/two stage: SSD; HAD | Longan without stone/color, shrinkage, and microstructure, drying temperature of SSD (120°C, 140°C, 160°C, 180°C and HAD temperature 60°C, 70°C | • SSD + HAD using 180°C SS (drying time 0.25 h), followed by 70°C HA (drying time 13 h) was the most suitable drying condition.<br>• Mathematical model for drying rate and convective heat transfer coefficient was validated. |
| 3. | Namsanguan et al. [71] | Experimental/lab scale (SSD/HPD) and (SSD/ HAD) | Shrimp/shrinkage, color, rehydration behavior, texture (toughness and hardness), and microstructure | • SSD/HPD dried shrimp had much lower degree of shrinkage, higher degree of rehydration, better color, less tough and softer, and more porous than single-stage SSD.<br>• SSD/AD gave redder shrimp compared to shrimp dried in a single-stage superheated steam dryer but no improvement was observed in terms of shrinkage and rehydration behavior. |

(*Continued*)

TABLE 3.3 (*Continued*)
Summary of the SSD Applications for Biological Products

G. Microbiological Inactivation and SSD

| SN | Paper Details with Authors | Type of Work/Dryer Type | Product/Quality Parameters | Remark |
|---|---|---|---|---|
| 1. | Park et al. [119] | Experimental on lab scale/ analytical modeling of the inactivation of geobacillus stearothermophilus spores in flour on different surfaces | Wheat flour Spore inactivation | • Spores during superheated steam treatment depends on heat and moisture transfer exchange between SS and different surface materials.<br>• Among the materials, the stainless-steel surface is better for the same, due to its higher thermal diffusivity.<br>• In addition to drying, sanitation of food can be studied as SSD works better for it. |

# 4 Lab/Pilot Scale or Commercial SSD Systems

## 4.1 INTRODUCTION

There are hundreds of types of dryers used in diverse applications using hot air and other non-conventional drying techniques. Even though, principally, every dryer operating on the hot air can be possible to convert with superheated steam, there are still engineering constraints and changes in the drying physics, which need to be understood for the conversion of HAD to SSD. Day by day, the use of SSD is increasing and now diverse types of dryers are developed for good number of products to be dried. It is possible to convert any hot air-based fluidized bed dryer to operate on SS media. In practice, major challenges while converting any HAD to SSD are the condensation of the steam at two important locations of material handling i.e. (i) supply and (ii) extraction process of product. Currently, drying chamber is surrounded with heat jacket/heater/steam tube, which almost eliminates condensation of steam in it. But for supply of feedstock and extraction of dried products, the special arrangement is required for continuous as well as batch operation. In this chapter, some of dryers working on the superheated steam and at lab, pilot or commercialized are summarized in the form of case study, for more detail understanding on them, reader may refer cited references appropriately.

## 4.2 SUPERHEATED STEAM FLUIDIZED BED DRYERS

Fluidized bed dryers (FBDs) are popular due to very high heat transfer capability resulting in higher drying rate and thermal efficiency, easy control, and transport of material. On the other side, there are issues like high electrical energy consumption due to requirement of high pressure drop, erosion of chamber and pipes, particles pulverization or attrition, etc., and refer "Applications for Fluidized Bed Drying" chapter by Mujumdar and Devahastin [120] for more details. The major components are fluidization chamber, supply and extraction mechanism, steam generator, and supply pipes. It is generally used for granular, powder, or materials which can be fluidized having size 10 μm to 10 mm such as coal, polypropylene, bagasse, pepper and coriander seeds, paddy, soybean, etc. It can be used in several industries such as pharmaceutical, dairy, food, chemical, dyes, metallurgical, etc. Details are discussed with help of the case study in the following sub-sections.

DOI: 10.1201/9781003275299-4

## Case Study 4.1: Superheated Steam Fluidized Bed Dryers for Low-Rank Coal Drying [121]

The continuous feed fluidized bed drier operating on superheated steam was initially designed based on air fluidized bed driers using American lignite. The initial scale of the thermo-gravimetric bed of approximately 10 g was extended to accommodate a bed mass of up to 5 kg. The main components are steam generator, superheater, fluidized bed with distributed plate, viewing port to observe the fluidized bed, screw feeder system for feeding the feedstock (coal), and cyclone separator with exhaust. The photograph in Figure 4.1 shows mainly the fluidized bed dryer along with controls and instrumentations.

The coal is dried by fluidization medium (steam) due to floatation in it providing thermal energy to raise the desired temperature and fed through the plenum chamber and distributor plate to fluidize bed. The coal is fed through the bed using screw feeder system connected to a variable speed motor to control the flow of coal. As wet coal is added, the height of the bed increases causing it to rise and overflow through the outlet and into the collection tray for analysis. Operating the fluidized bed at steady state over several hours allows steady-state moisture content to be reached, with the equilibrium moisture content of the specific conditions determined. Residence time was calculated by the volumetric flow into the vessel and overall volume of the bed. Various data is recorded during the drying such as temperature of the supply and exhaust steam, drying material at various locations in the fluidized bed; velocity of

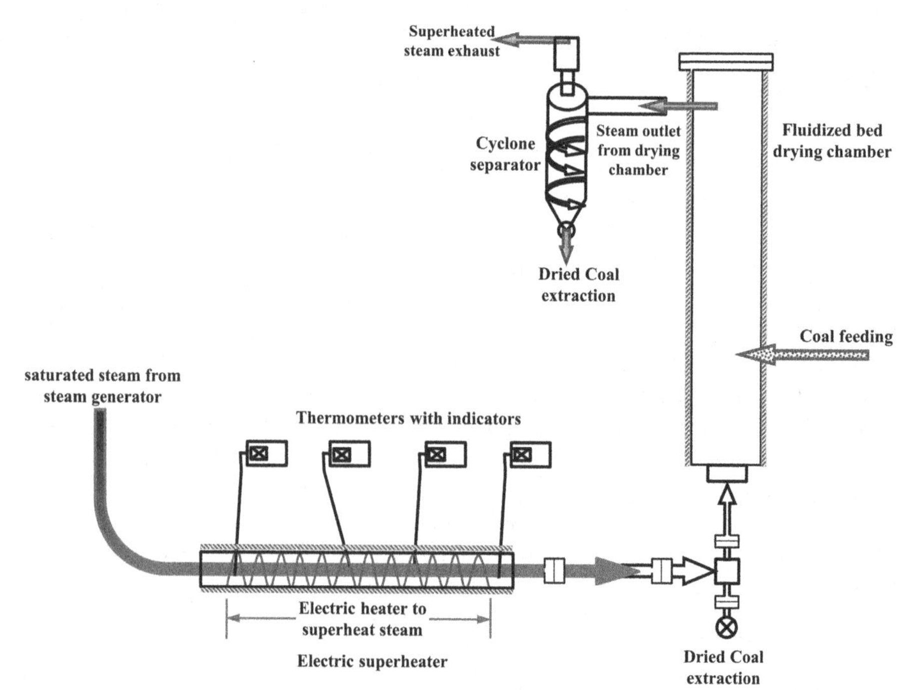


FIGURE 4.1 Continuous feed fluidized bed drier [121].

steam and coal, moisture removal, moisture content in the coal particle during the fluidized drying process.

The drying kinetics and consequences of the various parameters on the moisture removal of the coal are given in the following section. In this, the fluidization velocity, effect of initial particle size, and bed temperature with drying times on the dryer performance are investigated for Victorian brown coal. First discussion is given for batch drying of coal in FBD using SS and experimental results are used to compare an influence of scaling on the FBD kinetics. To maintain parity with the smaller experiments the 1 kg bed is comprised of completely raw coal. A drawback of this approach is that the fluidization characteristics in the early phase are undesirable and create a problem in initial bed sampling due to a lack of particle mixing within the bed. However, the results still provide enough resolution to accurately determine the drying kinetics. The details of the drying parameters with the type of the coal (Loy Yang, Yallourn) are given in Table 4.1 and the variation of coal moisture content with time is given in Figure 4.2 [121].

For SS at 130°C, there is visible condensation of steam in the initial heat-up period (also called warming period), but not for SS at 170°C. This may be because of the sudden increase in the feedstock temperature above condensation point due to higher temperature of SS. The drying time to reach the equilibrium moisture is 20 min for 170°C as against 30 min for 130°C superheated steam temperature, indicating the significant influence of temperature on the drying time. The drying kinetics for continuous dryer is given in the next section.

Continuous Drying Kinetics: Using a larger apparatus requires the initial bed to be dry rather than wet. To avoid de-fluidization, initially, a dried bed of coal is used to promote fluidization and then wet coal is fed into the bed, increasing the bed height, which causes the bed to overflow from outlet, where it is collected and analyzed. Once the outlet moisture content becomes consistent, this value is determined to be the steady-state outlet moisture content. The trend of moisture removal rate shown in Figure 4.2 for batch drying remains the same for continuous drying with higher bed material (3 kg). The SS temperature influences the equilibrium moisture content in dried coal at the outlet of the chamber. For the SS temperature increases (130°C–200°C), the stabilized outlet moisture decreases (ranges from 4% to 1.5%). It

**TABLE 4.1**
**Experimental Parameters for Superheated Steam Fluidized Bed Dryers**

| | Batch Drying | Continuous Drying |
|---|---|---|
| Variables | Parameters | Parameters |
| Coals | Loy Yang, Yallourn | Loy Yang, Yallourn, Morwell, Shenhua |
| Particle size (diameter) | 1.2–2.3 mm | 1.2–2.3 mm |
| Temperature | 130°C–170°C | 130°C–200°C |
| Velocity | ~0.45 m/s | ~0.45 m/s |
| Bed mass | 1 kg | 3 kg |

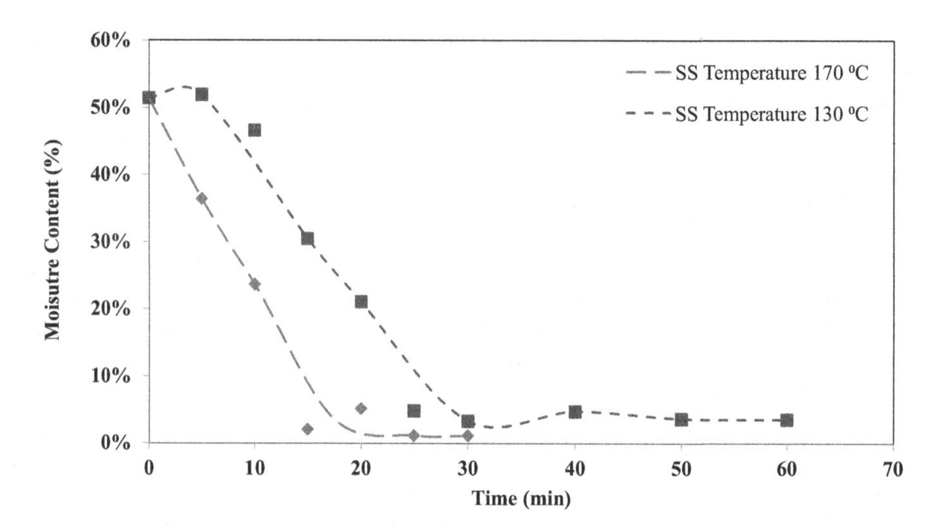


**FIGURE 4.2** Batch fluidized bed drying, 1 kg Yallourn brown coal [121].

is noteworthy to mention that the equilibrium moisture content is consistent regardless of the size of the equipment. Residence times shorter than 50 min for drying of all types of coal cause de-fluidization within the bed due to the pre-condensation resulting in increased wet coal mass. Overall, the initial fluidization medium temperature is the basis for the equilibrium moisture content, not the final fluidization medium temperature.

The lower SS temperature in fluidized dryer, the pre-condensation in heat-up period is significant on the surface of the coal in conjunction with a static feed point into the bed and density differences between dry and wet coal cause the wet coal to immediately stick together forming the lumps. At high feed rate, the wet coal instead of dispersing aggravates the formation of the lumps. This leads to increased density in the bed through which steam circumnavigates instead of breaking these cakes, resulting in slowing the drying rate and reducing the fluidizing within the bed.

**To overcome these following measures may be useful:**

- Pre-heat the inlet coal to remove any condensation forming and to reduce the coal density difference
    - As the de-fluidization has been partially attributed to the condensation across the particle, preheating through energy integration will increase the inlet coal temperature, preventing condensation and promoting bed mixing.
- Increase the drying temperature to prevent condensation from occurring
    - Another method of reducing condensation is to increase the inlet temperature of the fluidization medium. The caking effects were not observed at 170°C.
- Vary the coal inlet position

- If the wet coal spreads evenly across the bed the wet coal would not have the available mass to cause lumping and allow for a greater feed rate to be used.

For power generation, drying operation at lower temperatures can be the most energy efficient; therefore, initial heating of the coal has the greatest viability for increasing the throughput within the bed to reduce the lumps due to condensation of the steam when drying just begins.

**Physical characteristics of the coal:**

- The surface area and porosity show no change with the drying temperature, fluidization velocity, particle size, and drying medium.
- Particle breakage in coal is observed during the transition between bulk/bound water to non-freezable water. This occurs regardless of fluidization medium and drying method.
- Further attrition due to fluidization does not occur in a steam fluidized bed, but does occur in an air fluidized bed.
- The percentage change in the particle population during the attrition (in an air fluidized bed) is found to be linear, with the largest proportional change occurring to the larger particle sizes.
- Coal dried by superheated steam FBD re-adsorbs moisture lower than hot air fluidized or fixed bed dried coals (regardless of coal type).

---

### Case Study 4.2: SS Agitated Fluidized Bed Superheated Steam Dryers Test Rig [122]

Agitated fluidized bed superheated steam dryer (AFBSSD) system comprises five principal components, viz. the dryer vessel and integral agitator, the dried particle separator, the steam superheater, the recirculation steam blower, and the process controller as showing interrelationship of these various components in a process flow chart in Figure 4.3 [122] and schematic representation in Figure 4.4 [122]. The system is designed and developed with a complete array of safety interlocks so that it will work safely and unattended all the time.

---

**The objectives of the bench-scale test facility are:**

- Characterization of the dried particles in the steam atmosphere (e.g., particle size distribution, bulk density, odor, color, pathogens)
- Handling characteristics of the material before and after the steam dryer (e.g., stickiness, adhesion to the dryer wall)
- Heat transfer coefficients essential to the design of the full-scale dryer

The bench-scale system (Figures 4.3 and 4.4) consisted of a superheated steam source, agitated fluidized bed drying chamber, feedstock feeding system and dried product outlet system, cyclone separator, steam flow controller, temperature instrumentation on the walls

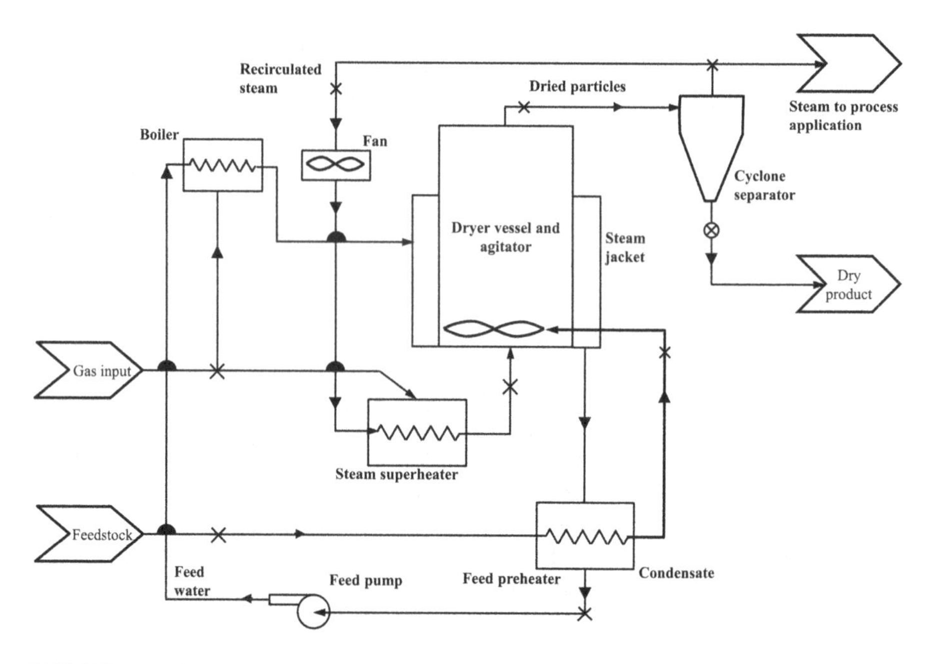


FIGURE 4.3 Schematic representation of SS agitated fluidized bed dryer (SSAFBD) at atmospheric pressure [122].

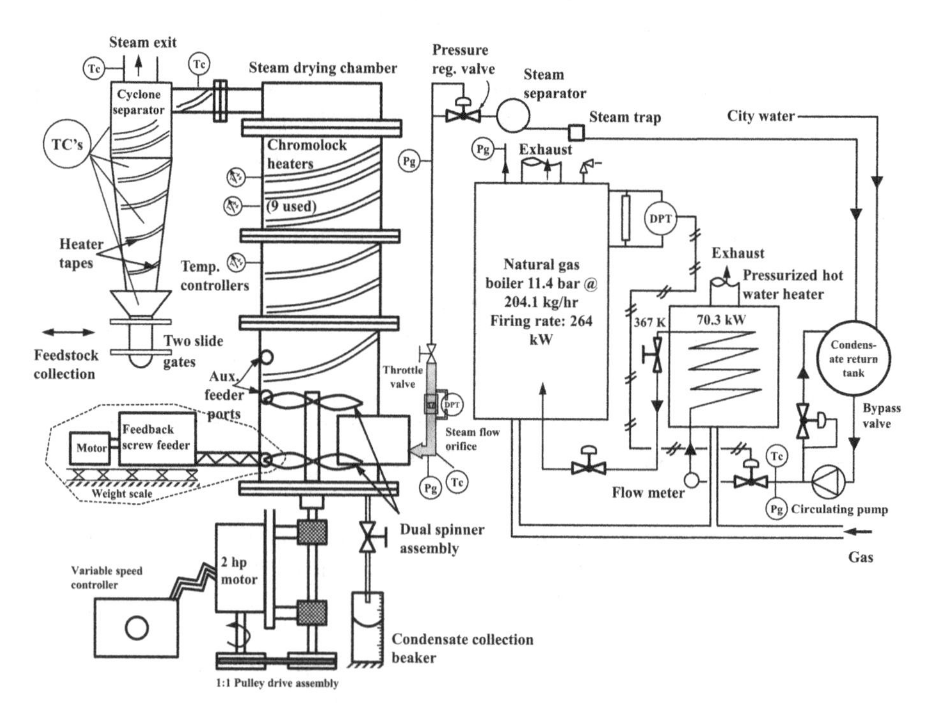


FIGURE 4.4 Schematic representation of bench-scale SS agitated fluidized bed dryer (SSAFBD) [122].

of the dryer and cyclone separator as well as within the dryer vessel, pressure gauges, and complete control systems for automatic and safe operation with data collection and storage. A PC-based data acquisition system was designed to continuously collect and store the temperature and steam flow rate data from the steam dryer experiments.

The dryer consisted of an insulated cylinder 0.2032 m in diameter and 1.07 m long. To minimize the sticking of the particles inside the cylinder, Teflon coating is provided on the inner layers of the cylinder. Generally, in the prototype, the dryer vessel is jacketed to promote the indirect heating and drying of the wet particles, enhancing wall-to-particle and wall-to-steam heat transfer coefficients but, in this model, both the drying chamber and cyclone separator were electrically heat-traced to simulate wall heating for simplicity of manufacturing, operation, and control. At the bottom of the vessel was a variable-speed agitator for breaking up the wet material and enhancing the vortex field. The agitator served to fluidize the particles as they entered the bottom of the dryer. The top of the dryer incorporated a vortex finder to retain wet material while allowing dry material to pass through to the cyclone separator.

Water from the city main was preheated in a gas-fired heater and then further heated to generate steam in the pressurized steam boiler. The steam is throttled from pressure approx. 5.17 bar to a little higher than 1 atmospheric pressure (1.034 bar) at the flow rate capacity of approx. 317.5 kg/h is throttled appropriately to get the superheated steam. The steam leaving the cyclone separator was vented to the atmosphere rather than reheated and returned to the dryer, as would be the case in the full-scale prototype system. Therefore, the bench dryer operated in an open-loop without energy and steam recovery. A 0.00635 m diameter tube had injected the slurry into the dryer vessel at the same level as the internal paddled spinner. The spinner effectively atomized the slurry mixture as it entered the dryer, as shown in the schematic of the slurry injection system apparatus (Figure 4.5). The superheated steam

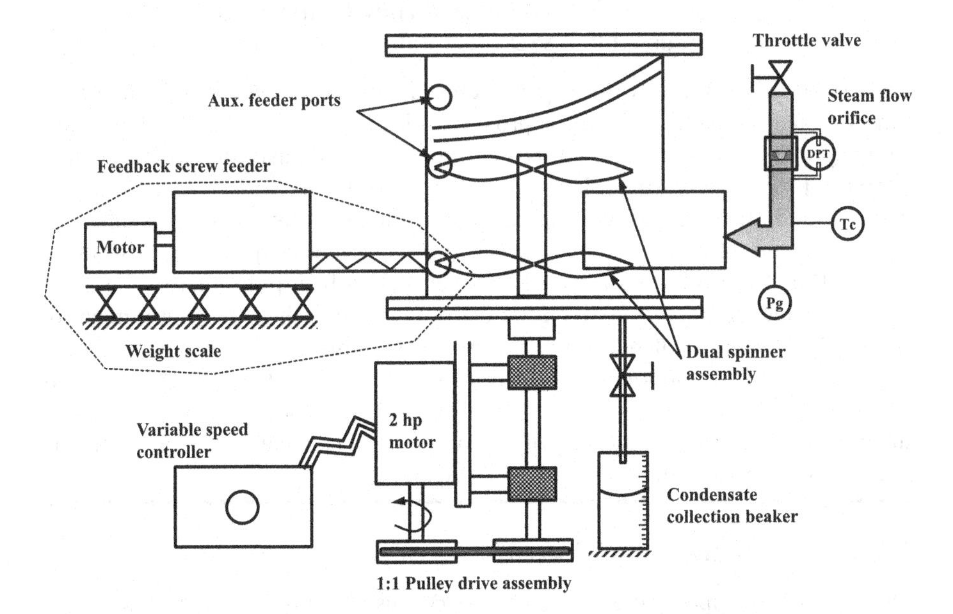


**FIGURE 4.5** Feedstock screw feeder system to drying chamber with variable speed controller [122].

at appropriate pressure and temperature (a little above atmosphere) is supplied to the drying chamber, fluidizing the feedstock, during which heat exchange takes place between the slurry (feedstock) and superheated steam, resulting in the drying of the slurry. The steam at a lower temperature will be exhausted into the atmosphere from the exit port located at the top of the cyclone separator and an equivalent amount is supplied so that fluidization will be maintained. The two gate valves located at the bottom of the cyclone separator worked very well to allow the collection of dried powder samples while allowing the testing to proceed uninterrupted.

The wet feedstock tested in the dryer apparatus consisted of a kaolin clay material in slurry and wet-cake form and municipal sludge material from three different facilities, in Albany and Syracuse, NY, and Boston, MA, United States. The sludge from the Massachusetts Water Resource Authority viz., Boston and Syracuse facilities were anaerobically digested and mechanically dewatered, and the Albany sludge was not digested but just dewatered. After completion of the kaolin tests, municipal sludge samples were dried in the system. Samples of the clay and sludge before and after drying were analyzed for particle size distribution, bulk density, color, residual odor, nitrogen and ammonia content, and volatile organic solids.

Kaolin Clay Feedstock: A kaolin clay product was mixed to a 50% by dry weight water content such that the mixture is a pumpable slurry. After the dryer had attained steady-state conditions, the slurry was pumped into the steam dryer using a peristaltic pump. The test with this was successful. The clay powder collected from the cyclone was identical in appearance to the dry clay product originally used to mix the slurry. There was no discoloration or contamination from the dryer apparatus or from the steam-heating at the elevated temperatures.

---

**Case Study 4.3: Pressurized Fully Commercial Fluidized Bed SSD (Beet Pulp Drying) [30]**

Traditional drum dryers with hot air are used for drying beet pulp, which consumes a huge amount of energy approx. one-third of the total energy necessary by a sugar factory to dry the entire beet pulp. The exhaust emitted from drum dryers is prone to considerable air pollution, mainly the emission of solid particulates and stinking gases. Over the years, for the conventional systems, the energy consumption in the sugar beet factory without pulp drying is reduced from 300 to 180 kWh/ton of beet. Moreover, to dry beet pulp, there is an additional demand for energy 90 kWh/ton of beets, making the total requirement of energy 270 kWh/ton of beet with traditional systems. In an average size beet sugar factory of 14,000 tons of beet per day, it uses 157.5 MW as fuel which corresponds to 542 tons of coal/day. One-third of this can be saved by pressurized steam drying of the pulp instead of drum drying. The steam dryer designed by EnerDry is discussed in detail in the following para:

---

### 4.2.1 Construction and Working

The pressurized superheated steam (3–4 bar) is used to fluidize the beet pulp and transfer the thermal energy required for drying. The important components of the dryers are the drying chamber, heat exchanger, dust separation, and steam circulation fan (see Figure 4.6). As presented in Figure 4.6, the beet pulp conveys using the rotary valve (1) and screw feeder (2) into the pressurized drying vessel (3) having ~4 bar superheated steam surrounded by heat exchanger (12). An impeller (4) is the only moving part in the dryer, circulating this superheated steam up through the perforated curved bottom (5) into a low, ring-shaped fluidized bed (6) fluidizing the beet pulp and swirling around as the arrows show. Guiding plates (not seen in figures) direct the pulp, and superheated steam moves them forward in the next cell of the ring-shaped fluidized bed around the heat exchanger (12) such that from the last cell they are taken out appropriately by discharge screw

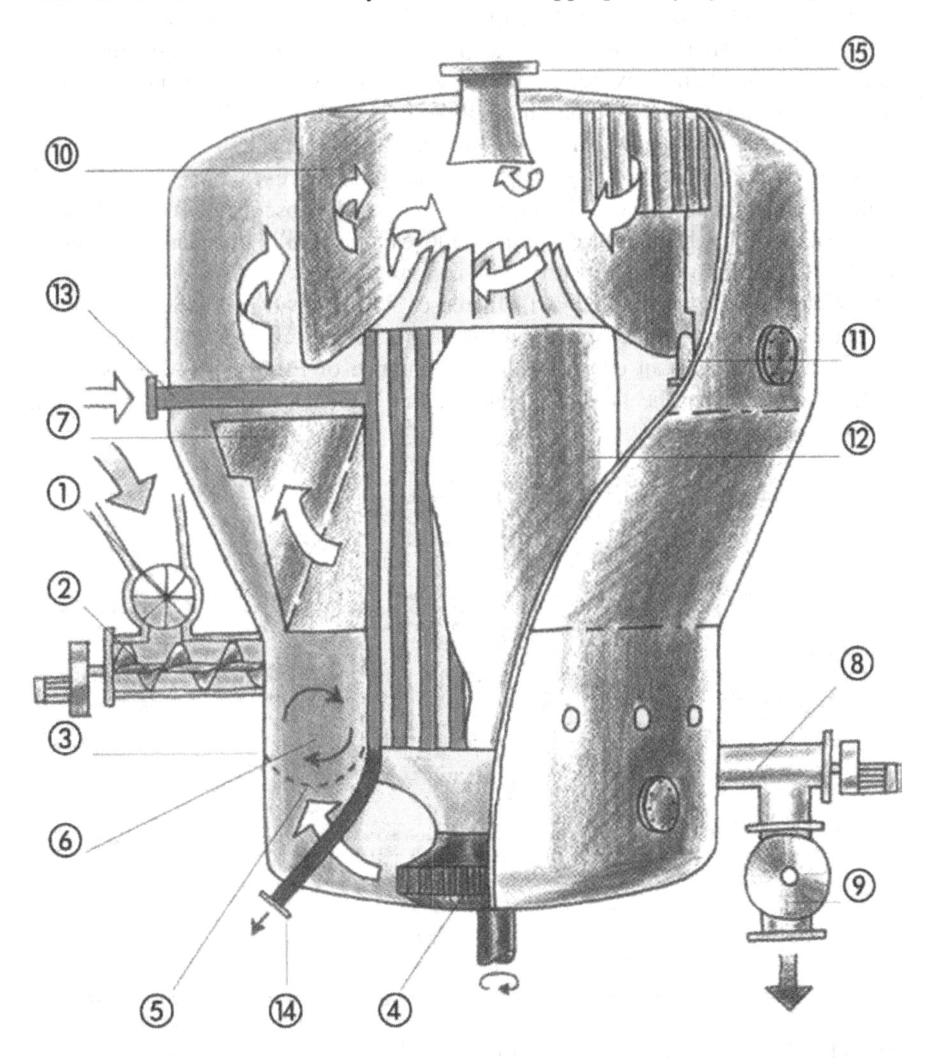


**FIGURE 4.6** The schematic diagram of industrial-scale high-pressure SS fluidized bed dryer [30]. (a) Capacity of driers with pressure of the superheated steam supplied [30] and (b) energy and mass balance with various site data [30].

(8), and rotary valve (9). Lighter particles are lifted up and reached between the plates (7) due to the flow of the superheated steam. These inclined plates allocate the superheated heated steam flow in the large cross-section, reducing its velocity. Because of the lower velocity, the light particles fall onto the forward inclined plates, slide down on those, and pass the gap between the plates and the conical vessel wall. Subsequently, the lighter particles also appropriately guided to move ahead in the next cell of the dryer and from the last cell they move toward the discharge screw (8), and pass the rotary valve (9) to get exit as dried product. The circulating drying media (superheated steam) reaches upper portion of the dryer, where separation of dust is performed in the main cyclone (10). An ejector takes out the dust collected in the cyclone through the pipe (11) and exits with the dried product. The dust-free superheated steam at lower temperature is reheated while passes down inside heat exchanger (12) by higher pressure steam using its mainly latent heat supplied through the pipes of the heat exchangers (13). Due to loss of latent heat, steam gets condensate, and leaves the dryer through the pipe (14).

With increasing the pressure of the supply steam to heat exchanger, a temperature of the circulating superheated steam is also increased, subsequently raises the drying capacity of it, when circulating fan (4) moves it upward in the ring-shaped fluidized bed chamber (6). Consequently, with increased supply pressure of the reheating steam, the capacity of the FBD will increase, as illustrated in Figure 4.7a for the moisture evaporation capacity of the different sizes of dryers.

Figure 4.7b shows an energy and mass balance of FBD, where a condensate outlet from the main heat exchanger located at the center is utilized to pre-heat

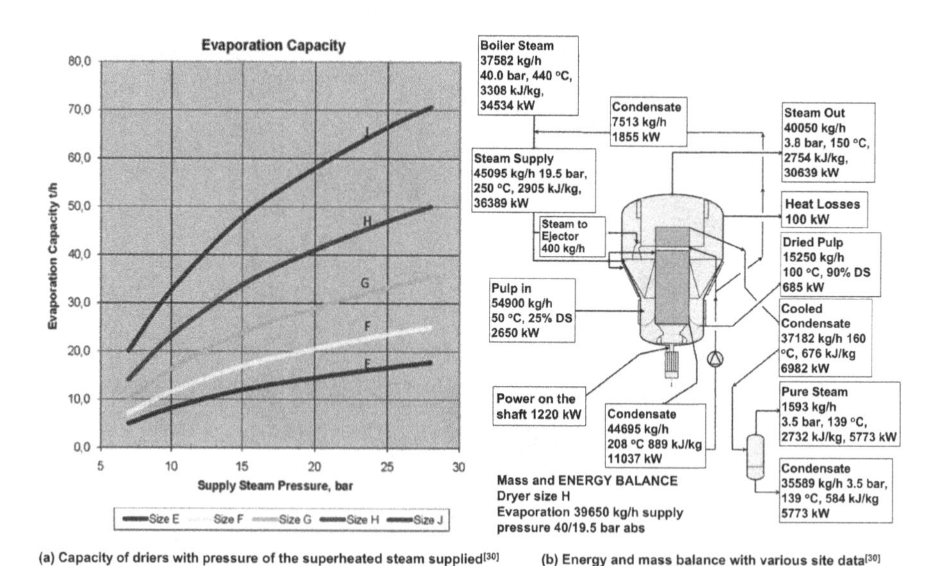


(a) Capacity of driers with pressure of the superheated steam supplied[30]

(b) Energy and mass balance with various site data[30]

**FIGURE 4.7** Sizing and mass-energy balance of the high-pressure superheated steam fluidized bed dryer [30]. (i) Hot air drum drying of beet pulp and (ii) steam drying [123]. (a) Energy flow diagram for sugar factory with operation from November 2013 [30]. (b) The dryer in Nakskov, Denmark.

recirculating superheated steam in an additional heat exchanger placed above the main heat exchanger. That drops the amount of supply steam at high pressure to the main heat exchanger, and the amount of superheated steam exits from the dryer and the pure steam from the flash tank is even marginally greater than the supply superheated steam approximately the same as the amount of the moisture evaporated. As shown in Figure 4.7b, mass, temperature, pressure, enthalpy, etc., are given for steam and beet pulp at various locations of the novel industrial dryer. The electrical energy used for the fan is recovered in the dryer in the form of thermal energy, increasing the enthalpy of the steam leaving.

All the energy supplied to the dryer utilized for steam heating, which is in the case of the sugar factory in the evaporation plant (dryer) for juice concentration. That means the drying takes place without any use of energy. The borrowed energy is fully returned and used; furthermore, the severe pollution from the classic drum dryer is fully avoided. The steam from the dryer can also be mechanically compressed and used as supply steam for the dryer itself. In this case, the needed mechanical/electrical energy will be about 14% of the energy needed by a drum dryer. It is like a heat pump dryer with a ratio of 1:7.

Figure 4.8a compares the energy flow through a sugar factory with drum drying of the beet pulp with a factory with steam drying of the pulp. From this example, it is seen that the 40.2 MW fuel used for the drum drying is fully saved. Furthermore, the steam boiler house uses 82.5 − 77.3 = 4.2 MW less fuel. Less steam goes through the turbine in the sugar factory's own power plant, the power production will go down by 3.9 $MW_e$.

The first 7 years of the development of the pressurized drying technology began after the energy crises in 1981. It was realized it would be an enormous advantage if it would be possible to dry the beet pulp in a closed vessel under pressure in its own vapor. Then it would be possible to get from the pulp out as 100% usable steam. There

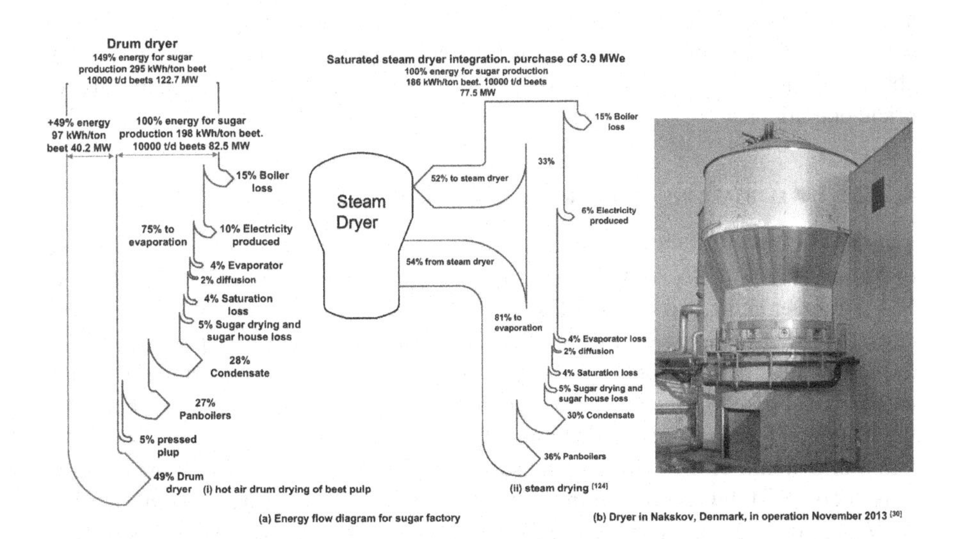


**FIGURE 4.8** (a) Energy flow diagram for HA drum dryer and SSFBD. (b) Photograph of the industrial-scale high-pressure SS fluidized bed dryer [30,123]

was no useable basic info found in the literature at that time. A team under the leadership of Arne Sloth Jensen worked in the laboratory in Denmark for 2 years on fundamentals like finding drying time, reaction in the product under the given conditions, which drying process should be used in the closed vessel, quality of condensate, heat transfer in the fluid bed, drying time as a function of pressure. Some of the results were:

- The pressure shall be high enough for the purpose the steam from the drying chamber shall be used for heat source in the same process.
- Higher pressure in the drying chamber gives faster drying. It looks like the increased density increases the drying speed similar to increase of the heat transfer by gas density.
- Higher pressure saves power on the fan for internal circulation.
- Higher pressure reduces the size of the apparatus.
- Drying under pressure changes cellulose to hemicellulose, which improves the digestibility of beet pulp used for cattle feed.

The negative effect of the pressure:

- The wet particles will have a higher temperature during the drying. That prevents many products from steam drying under pressure. Maillard reaction will take place in beet pulp at 3.7 bar, so it must be out of the dryer after 5 min, which fits with the drying time.
- It is difficult to get a product in and out. A complicated rotary valve has been developed.

Other feedstock:

Distillers grain from ethanol production, mash from beer brewing, solids left out after fruit juice production, remained solids after production of starch from corn or wheat, wood chips, brown coal, sawdust, bark, bagasse, etc. application can be suitable for the above dryers, but parameters need to be optimized for efficient and appropriate operations.

### 4.2.2 Investigations of Energy and Carbon Footprint

Two largest dryers of size "J" shown in Figure 4.6 are developed; one is installed in Nampa, Idaho, USA, and another, the same size dryer, operating successfully at Nakskov, Denmark, from 2013 (Figure 4.8b). Out of these two, the former steam dryer (installed at Nampa) substituted three coal-fired drum driers operational from the 1950s. For these old coal-fired dryers, major maintenance and reduction of the local environmental impact due to carbon emission were critical issues. Therefore, the demands from the environmental authorities were accomplished by the installation of a steam dryer, which at the same time could give a daily savings on 200 tons of coal and thus reduction in the emission of $CO_2$ by approx. 600 tons per day.

Drying in steam under pressure for beet pulp and similar feedstock has the following advantages other than ones mentioned in Chapter 1 over conventional hot air dryer:

- Nearly 100% energy saving and thereby a large $CO_2$ emission is avoided.
- The air pollution from convention drum drying is fully avoided.
- No product losses due to burning as happens by drum drying.

### 4.2.3 Lab-Scale Continuous Agitation Fluidized Bed SS Dryer

After the successful completion of the tests on bench-scale dryer, lab-scale system with a 113 kg/h feed rate system is designed and fabricated on which sludge drying tests are carried out. The additional facilities provided are a closed-loop steam circulator (or blower), steam superheater, packaged steam boiler, and sludge-handling system.

In operation, a spinner/agitator placed at the bottom of the vessel broke up mechanically the feedstock, mainly in the form of a cake or paste-type consistency, to form wet granules that entered the dryer. The mechanical spinner induced a vorticity in the steam atmosphere and thus helped promote the vessel's internal recirculation of steam currents. These internal circulation paths were essential for entraining the wet particles in the steam and retaining them within the small dryer vessel body until the particles were dried. Although the bottom spinner helped promote this vorticity field, the internal recirculation currents were primarily sustained by the tangentially directed carrier steam inlet flow into the dryer. However, the fluidization of the partially dried feedstock particles also helped promote conduction heat transfer among the particles and helped continually introduce new particles. When the material had been dried sufficiently, it exited from the top of the dryer and entered the cyclone separator. The dry material was removed at the bottom of the cyclone through a rotary steam lock.

For the steam atmosphere dryer, wet sludge alone was dried successfully and mixing of wet and dry sludge during feed found unnecessary. The SS and hot air (HA) dried sludge was found the same chemical and physical properties of the particle. In certain features, the SS-dried sludge shows superior quality as more granular and pelletized than HAD. Under the same operating conditions and drying up to 5%–7% wetness, SSD gives 25% higher performance than HAD.

Overall, wet, municipal solid waste as feedstock from an anaerobic digester was dried and converted to salable product, a high-nitrogen content fertilizer. The direct-contact superheated steam atmosphere agitation fluidized bed dryer was shown to be energy efficient saving more than 25% energy due to recirculation of steam over conventional HA dryers in addition to the energy recovery. Few more advantages are oxygen free drying, reduction of air effluent, reduced odor, and improved dried product quality. The dried product quality and quantity were found to be superior or at par with HAD.

---

**Case Study 4.4: Low-Pressure SS Fluidized Bed Dryer [87,111,124]**

The vacuum or low-pressure dryers with superheated steam have added advantage of drying of temperature-sensitive materials by reducing the operating pressure below atmosphere. This leads to the lowering of the saturated temperature below

100°C and can be maintained as per the requirement of the feedstock. Kozanglu et al. [87,111,124] developed lab-scale experimental setup of superheated steam vacuum fluidized bed dryer (SSVFBD)/low-pressure superheated steam fluidized bed dryer (LPSSFBD) and investigated the drying kinetics with product quality for various food products such as pepper seeds, paddy, and coriander, discussed in the following section. The general setup of experimental test rig was same for all the test materials.

#### 4.2.3.1 Experimental Setup

The important components of the test rig are the fluidization column, temperatures and pressure measuring probes fitted at various locations, orifice plate with U-tube manometers for flow measurement, rotary pump to establish vacuum in the chamber, steam generator, feedstock supply and dried material removal systems, etc. as shown in Figure 4.6. An electrical heater with 9kW was connected, supplying energy to the superheater before the column to provide superheated steam. The fluidization cylindrical section of 0.7m high and 0.1082m inside diameter manufactured from stainless steel has visual observation of the fluidization process by a set of holes (diameter 0.0254m) perforated on the walls of the column and covered with glass-forming windows. A stainless steel distributor of 0.002m thick was used to hold the drying material in the chamber and allow the superheated steam to fluidize during operation. Furthermore, variable electrical power resistance heaters mounted on the walls of the fluidization chamber provided additional heat to prevent condensation in it by improving the wall-to-particle heat transfer coefficient. During the operation, a temperature was maintained constant by a control loop linked to the resistances of heaters. For a given time, the column wall is also at the operating temperature. The absolute pressure also varied in stages (such as 40, 53, and 67kPa) to find the influence of an increasing vacuum on the drying kinetic and dried product quality. At downstream and upstream of the column near the bed, the operating pressures are measured by a vacuometer through a mercury differential manometer mounted at the top of the bed, respectively. To maintain the vacuum at various pressures, appropriate arrangements with vacuum pumps and valves were made. Three different types of feedstock used for drying in the test rig were: coriander seed ($d_p$=3480μm; $\rho_p$=620kg/m$^3$; $\Phi$=0.87; $C_i$=1.22–1.45); pepper seed ($d_p$=4500μm, $\Phi$=0.95, $\rho_p$=840kg/m$^3$; $C_i$=0.55–0.59); and paddy drying ($\rho_p$=1180kg/m$^3$, $d_p$=4570μm, $\Phi$=0.74). The drying kinetics and dried product quality will be explained next for the material tested on this dryer (Figure 4.9).

#### 4.2.3.2 Results and Discussion

The drying with SS at temperatures 90°C, 100°C, and 110°C, pressure 40kPa (ab), and velocity 2.35m/s was shown in Figure 4.7 for both coriander and pepper seeds. For a given temperature, the moisture content in the given seed is reducing at a faster rate initially due to constant drying rate and after critical moisture it slows down with time as shown in Figure 4.7. The supplementary heat provided by the electric resistance heaters fitted on the external walls of the fluidized bed column helps to have temperature of feedstock higher than the saturation temperature of steam at chamber pressure and hence the initial moisture gain is not observed for coriander and pepper seeds for all drying temperatures (exception 90°C for pepper seeds, Figure 4.7);

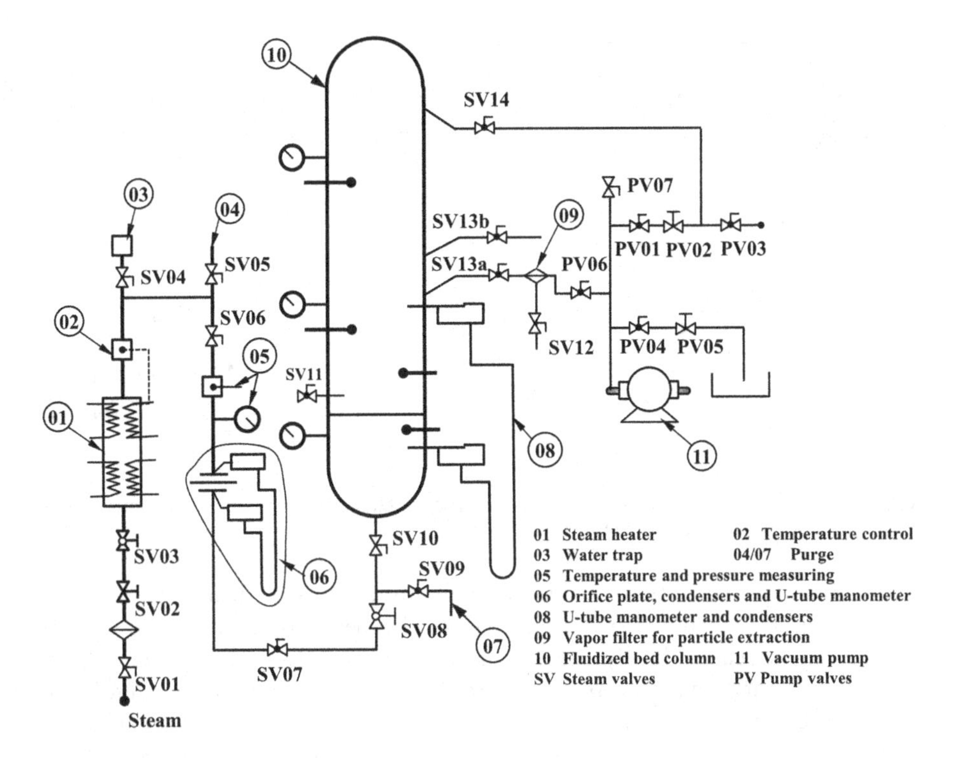


**FIGURE 4.9** Superheated steam vacuum fluidized bed dryer [87,111,124].

however, there was reduction in the moisture from the beginning of the drying process as shown in Figure 4.7 for coriander seeds only for all temperatures and for pepper seeds except 90°C condition. Further, higher temperature of SS corresponding to higher degree of superheating, $\Delta T = T_o - T_{sat}$, helping to accomplish a drying process without pre-condensation (exception 90°C for pepper seeds, Figure 4.7). With increasing the temperature (90°C, 100°C, and 110°C) of drying media, drying time reduces by significant margin (3000–1900 s for coriander and 3600–2500 s for pepper seeds) as shown in Figure 4.7. Further, the effect of the variation of pressure on the drying rate and time shows with decreasing the pressure, drying time reduces mainly due to increasing the degree of superheat. Further, it is important to note that for coriander the degree of superheat 2°C–4°C will have significant effect and with increasing it, no effect on final moisture contents (Figure 4.10).

## 4.3 SUPERHEATED STEAM ROTARY/DRUM DRYER (SSR/DD) [125]

Generally, in the rotary dryers, the feedstock is passed through one end of the cylindrical drum with drying media such as hot air/moist air and from the other end, dried feedstock (product) is taken out. The drum is rotated with some velocity so that the feedstock will have spiral movements along with heat transfer media, having a higher heat transfer rate. But when it comes to changing it to superheated steam as drying media, there are certain engineering issues, mainly leakage losses at the feeding

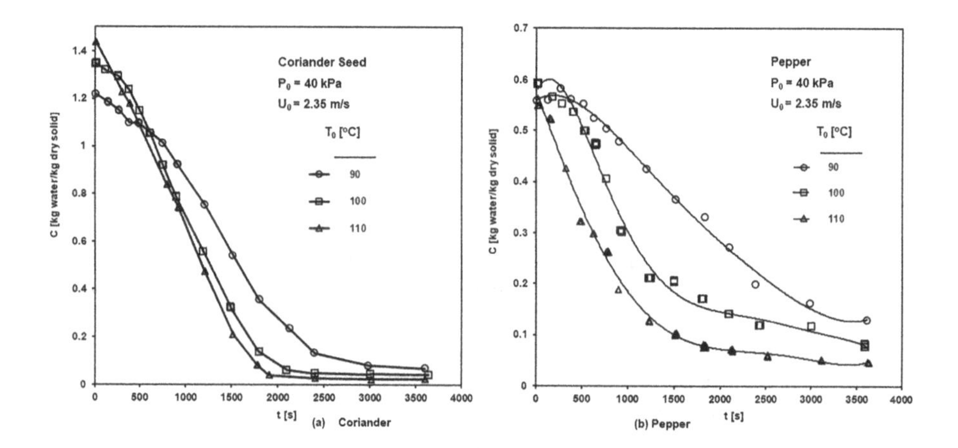


**FIGURE 4.10** Moisture content vs. time for (a) coriander and (b) Pepper at various superheated steam temperatures [87,111,124]

and extraction section. Further, due to different properties of steam than air, such as lower viscosity and higher thermal conductivity, heat capacity rate, heat transfer coefficient, etc., can definitely affect the heat and mass exchange in the rotary dryer.

The rotary dryer with superheated steam is designed and developed according to the engineering standards and codes as a co-current triple-pass for the demonstration purpose [125]. This can be used to dry a wide range of pasty and fibrous products and divided solid. It can handle products with a substantial size variation and is flexible in operating parameters such as a wide range of pressure, flow, and temperature conditions. First test was performed for pressed beet pulp at the industrial site, which received valuable inputs for further development.

### 4.3.1 Construction and Working

A representative sketch of the rotary dryer demonstrator with complete system integration is shown in Figure 4.8 [125]. It contains three subsystems: (i) dryer chamber for drying and solid-steam separation with supply and discharge of drying feedstock, (ii) steam generation with superheating, and (iii) energy recovery and recirculation unit. As discussed the triple-pass rotary dryer is developed by Maguin Company, France.

In this, the drying process takes place triple-pass co-current with direct contact between the superheated steam and the feedstock as shown in Figure 4.9. The three concentric drums are arranged as one central section and two annular sections with diameters 0.3, 0.52, and 0.8 m, respectively, and 2 m in length for all, forming three passes as shown in Figure 4.9. Generally, a divided solid is used as feedstock, first passes through the central section (first pass) and then the middle (second pass), and last outer annular section (third pass), completing three passes extracting the maximum possible energy from superheated steam. The feedstock is transported by means of the superheated steam flow. The light particles are passed with higher velocity than heavy/coarse particles due to lower density and size; hence the drying will be uniform irrespective of the particle size. While transferring the particle through the various passes, the interaction between the superheated steam and solid feedstock results in the exchange of energy and moisture, drying the particles (Figures 4.11 and 4.12).

From the exit of the outermost drum, the dried feedstock particles are passed through the cyclone separator, where dried product is separated and steam is exhausted by a rotary superheated steam lock valve. Due to the deflectors welded through the conical side of the cyclone, the residence time increases by forming anticyclone, which helps to recover the fine particles by backflow of the superheated steam toward the uppermost of the cyclone.

The heat recovery from the exhaust steam is performed by both partial recirculation of the low enthalpy and by condensation of the remaining steam. The recirculated steam same as supply steam is passed through the superheated heat exchanger by blower to raise its temperature to supply temperature so that it can used again for drying. This will definitely save the energy as well as amount of the steam. Further, the remaining exhaust steam, equivalent to the evaporated moisture from the feedstock, is transferred to cooling unit for recovery of thermal energy along with the amount of water by cooling it up to saturation temperature. The wet scrubber cleans the steam before passing to cooling unit. The non-condensable gases are removed and burned off. The superheater is constituted of six shell-and-tube heat exchangers, with the first two having 50 tubes (each tube dimensions: 21.3 mm diameter and 1 m length) and the remaining having 28 tubes of each diameter 26.9 mm, 1 m length. The superheated steam is passed through the tubes and hot gases surround the tube in shell-and-tube heat exchanger.

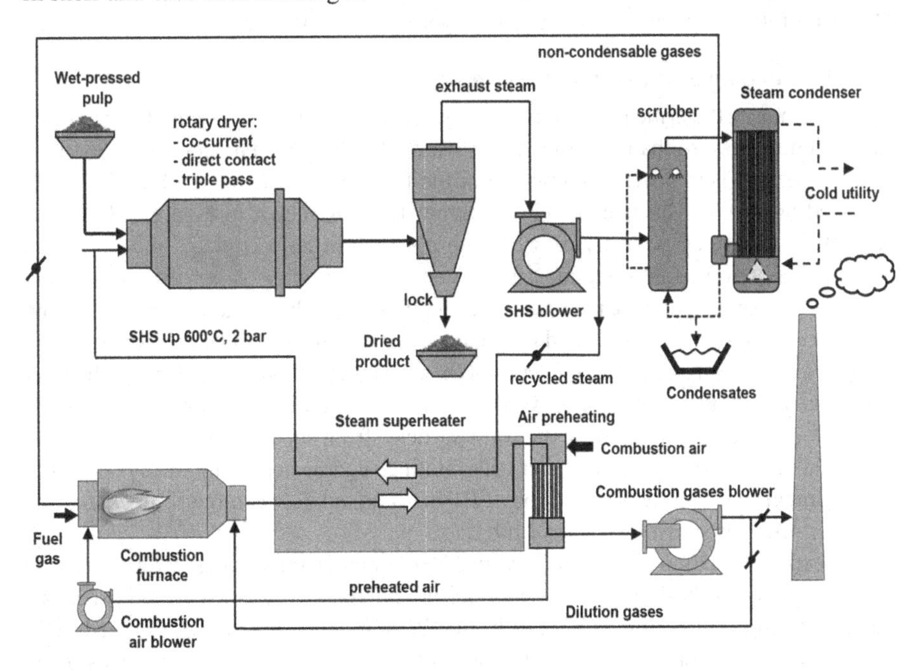


FIGURE 4.11 Overview of the drying process [125].

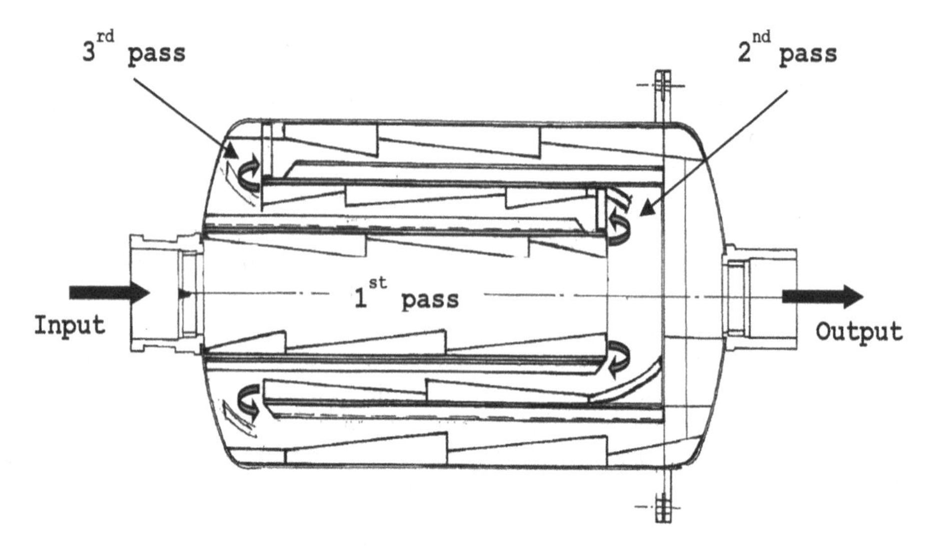


**FIGURE 4.12** Detailed view of triple-pass rotary dryer [125].

The dryer is operated slightly above atmospheric pressure (1–2 bar ab), so that air infiltration will be avoided and further helps for better heat recovery from exhausted steam. However, it requires better sealing of superheated steam at the feeding and extraction systems of feedstock, which may be provided by additional mechanisms. The evaporation capacity of this type of dryer will be 100 kg/h.

#### 4.3.1.1 Feeding and Extraction Systems

These are one of the critical and important systems in superheated steam dryer to operate at higher pressure than atmospheric pressure (1–2 bar ab). The design criterion for feeding systems is to provide a flow rate of the feedstock in the range of 42–145 kg/h. The feeding mechanism is consistent of three steps, as shown in Figure 4.10a and b viz.: (i) the feedstock is firmly pressed with a screw feeder; (ii) then, this compacted feedstock is conveyed in a retention tube to ensure the sealing as per the requirement between 1–2 bar (ab); and the retention tube length will be adjusted according to the type of the feedstock such as fibrous and granular; (iii) at the end, the wet feedstock is passed through a paddle screw conveyor to control the constant feed rate to the drying chamber maintained at steam temperature. After feedstock is dried in the cylinder, it is passed to a cyclone separator for settling the solid particles from superheated steam flow. This separated dried feedstock is taken out with rotating superheated steam lock. For further improvement in the outlet seal, a tube of 0.3 $m^3$ is added between the SS-lock and the conveyor screw as shown in Figure 4.10b.

The experimentation with beet pulp with 27% dry matter as feedstock at 240°C–300°C temperature and 1.09–1.24 bar (ab) pressure of superheated steam was dried successfully, resolving initial hurdles mainly in the feeding and extraction of the feedstock. These can be overcome appropriately and after that longer duration test needs to be performed for drying characterization and parameter optimization (Figure 4.13).

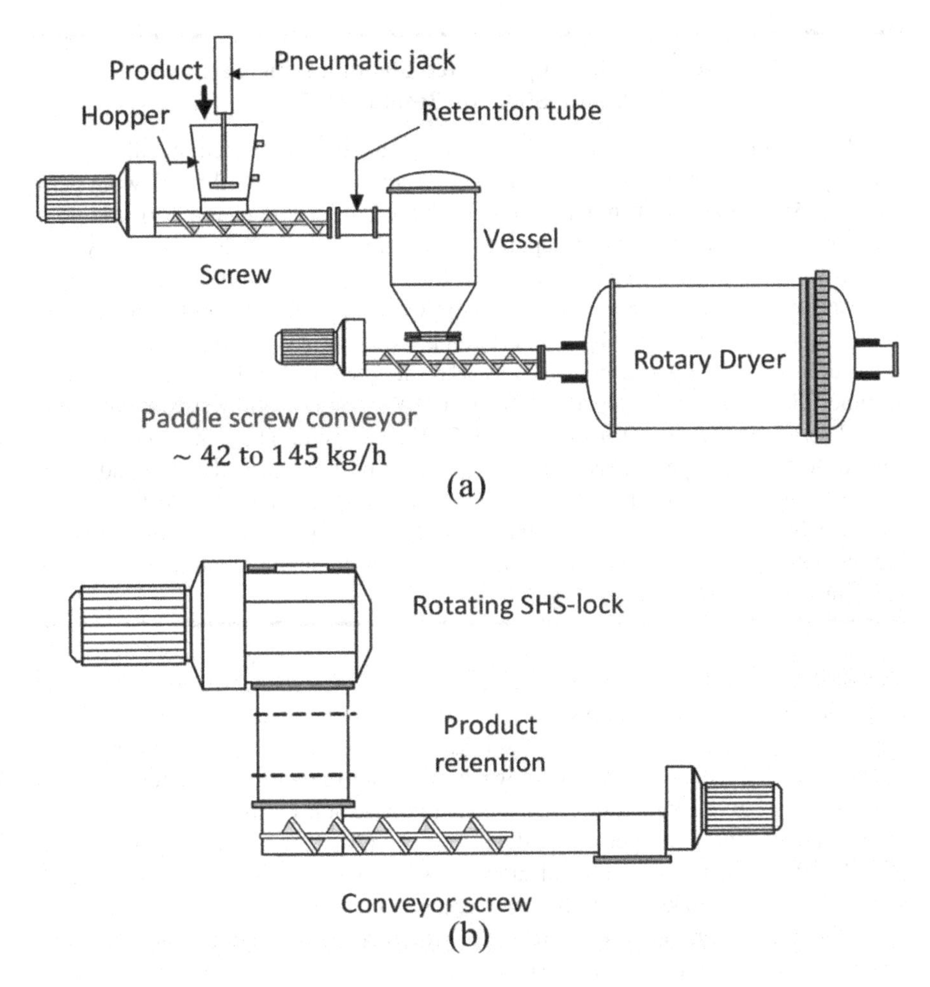


**FIGURE 4.13** Product (a) feeding and (b) extracting systems [125].

## 4.4 SUPERHEATED STEAM IMPINGEMENT DRYER (SSID)

Impinging dryers are generally more suitable for the materials which are in the form of thin sheets or require a conveyor for movements, such as papers, photographic films, textiles, fabric (woven/nonwovens, carpets, etc.), lumbers, fruits and vegetable slices, etc. In the course of impingement drying, the high velocity and at appropriate temperature heating media (HA, superheated steam, etc.) are impacting on the surfaces of the products. The floating particles are creating pseudofluidized bed with hot air/superheated steam in it. Due to this, the thermal boundary layers thickness reduces significantly, raising the rate of heat and mass transfer by many-folds compared to convection heat transfer. Due to this, either drying time will be lower or drying can take place at lower temperature than the conventional convection drying. The important applications and future trends mainly for performance improvement and potential technologies are discussed by Xiao and Mujumdar [126] especially for food products.

### Case Study 4.5: Superheated Steam Impingement Dryer (SSID) Soy Residue [105]

The line diagram of lab-scale test setup for coaxial superheated steam dryer of two-impinging streams is shown in Figure 4.11 and discussed in the following paragraph [105]. It consists of important components like, drying chamber (made up of stainless steel), steam generator, superheater, feeding and extraction system, cyclone separator, various measurement and control devices (temperatures, pressure, moisture, and flow rates), etc. Steam generator produces saturated steam having a maximum mass flow rate of 500 kg/h at a steam pressure of 7 bar (gauge) of water-tube configuration, which was used to supply it to the system. The drying chamber with an inner diameter of 0.25 m and volume of 0.018 $m^3$ has two inlet streams of superheated steam impinging at a distance of 5, 9, and 13 cm. The exhausted steam is passed through the cyclone separator, where dried particles are taken out and steam is filtered by a stainless steel screen (200 holes per square inch) to further remove the particles from the steam. This exhaust steam is recirculated by a high-pressure blower with a rated capacity 2.2 kW after reheating to supply conditions appropriately into the drying chamber (Figure 4.14).

The drying is performed on soy residue (okara), a test material, generated as a by-product of a soymilk process, having initial moisture content 82%–85% (w.b.) or 4.5–5.6 $kg_{water}/kg_{dry\ solid}$. Before drying began, the soy residue was dewatered by a hydraulic press, removing the excess water, reducing the moisture content to 72%–75% (w.b.) or 2.6–3.1 $kg_{water}/kg_{dry\ solid}$. The parametric study was performed by varying parameters such as superheated steam velocity (20, 27 m/s), temperature (130°C, 150°C, 170°C, 190°C), impinging distance between two pipes (5, 9, 13 cm), dried feedstock flow rate (10, 20 $kg_{dry}$ solid/h). Another parameter considered is the recirculation ratio of steam, which increases with increasing the steam temperature, however for a given velocity, recirculation of the steam is remains constant mainly due to the constraint of the size of the pipe and no other control mechanism is available. Volumetric water evaporation rate strongly depends on the temperature for both steam velocities and impinging distance for lower steam velocity. With temperature increases, the volumetric evaporation rate also increases by a significant margin for both velocities, but with increasing impinging distance, there is a significant reduction in volumetric evaporation rate at the lower velocity of SS (20 m/s) but at higher SS velocity (27 m/s) reduction is by a small margin. The specific energy consumption of the high-pressure blower ($SEC_{blower}$) and electric heaters ($SEC_{heater}$) reduced with rise in the inlet steam velocity, inlet steam temperature, and material feed flow rate. It is encouraging to note that the savings in total specific energy consumption of SS compared to HAISD with and without exhaust air recycle were to the tune of 46% and 95%, respectively.

## 4.5 CLOSURE

Superheated steam drying is having wide applications and day by day researchers are trying hard to use it in several diversified areas. In this chapter, a few lab-scale,

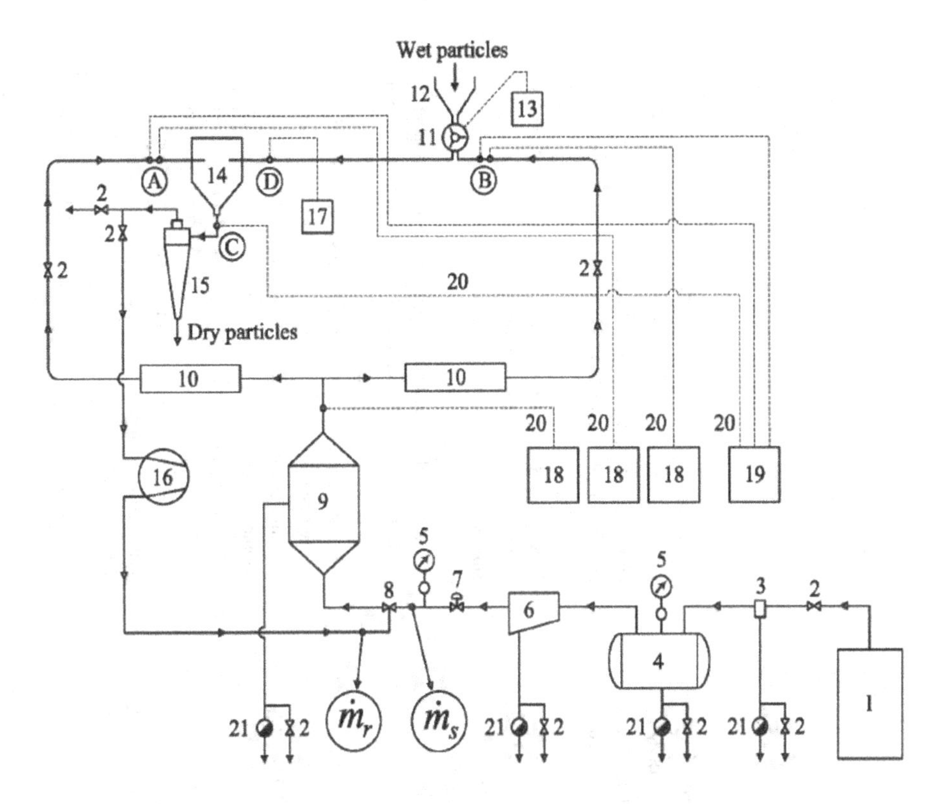


**FIGURE 4.14** A process flow diagram of superheated steam impingement dryer (SSID) with associated units [105]. (1) Boiler, (2) globe valves, (3) steam pocket, (4) steam header, (5) pressure gauges, (6) steam separator, (7) pressure-reducing valve, (8) steam ejector, (9) steam superheater, (10) supplementary electric heaters, (11) star feeder and DC electric motor, (12) hopper, (13) voltage regulator, (14) drying chamber, (15) cyclone, (16) high-pressure blower, (17) velocity measurement probe, (18) Propotional Integral Derivative controllers, (19) temperature data logger, (20) thermocouples, and (21) steam traps.

pilot, and commercially available systems are discussed, with a particular example taken as a case study. The three case studies on FBDs discussed in this chapter are summarized as follows: (i) simple FBD on low-rank coal; (ii) continuous agitated FBD on kaolin clay in slurry and wet-cake form, and municipal sludge with different pretreatment; (iii) low-pressure FBD on seeds of pepper and coriander, and paddy. Further, a case study on a rotary/drum dryer using superheated steam as a demonstrator plant was proposed for pressed beet pulp and may be used for soy residues and similar feedstocks. Lastly, a superheated steam impingement dryer (SSID) was proposed for various applications and discussed for soy residue drying as the case study in lab-scale systems. The capability of superheated steam dryers for diversified applications is reflected in these case studies. There is definitely lot of work to be performed, mainly pilot and commercial scale trials on diverse products, to increase the applicability of SSD (Table 4.2).

**TABLE 4.2**
**Details of the SSD with Applications and Features**
**(A) Superheated Steam Fluidized Bed Dryer (SSFBD)**

| S.N. | Paper Details with Authors | Type of Dryer | Feedstock/Type of work/Features | Remark |
|---|---|---|---|---|
| 1. | Takahashi et al. [127] | Lab scale-circulating SS fluidized bed dryer with an internal draft tube | Dilute slurry of silica with moisture and dry silica as bed material/to evaluate the feasibility and drying kinetics. | • Size distribution of bed particles did not change appreciably during the steady state of operation.<br>• Developed circulating SS fluidized bed dryer successfully operated.<br>• There are certain modifications related to design of the dryer for more flexibility and performance improvement. |
| 2. | Potter [24,128] | Pilot plant-SS fluidized bed dryer (SSFBD) | Brown coal (Victoria Australian)/pilot trial-effect of heat transfer surface, local pressure gradient, pressure fluctuations | • Summarized more than 15 years of work related to SSD of coal at Monash University Australia.<br>• The process of steam drying in fluidized beds is capable of drying high moisture material (like coal) efficiently and can be scaled up for industrial use. |
| 3. | Faber [129] | Lab scale SS fluidized bed test rig | Alumina and a highly hygroscopic molecular sieve—Comparison: FBHAD and FBSSD, Inversion temperature for equilibrium systems and non-equilibrium systems by mathematical modeling of the drying behavior in constant drying rate period, drying kinetics in falling drying rate region | • Below inversion temperature, for hygroscopic material (sieve), drying rate is significantly higher than non-hygroscopic (alumina) reducing overall drying time for SSD than HAD. Similar observation was reported for other products also.<br>• In SSD, energy costs are around 40% lower than in HAD, whereas capital costs are roughly 20% higher. The economic advantage is more apparent for the highly hygroscopic molecular sieve than for the weakly hygroscopic alumina for SSD. |

*(Continued)*

**TABLE 4.2 (*Continued*)**
**Details of the SSD with Applications and Features**
**(A) Superheated Steam Fluidized Bed Dryer (SSFBD)**

| S.N. | Paper Details with Authors | Type of Dryer | Feedstock/Type of work/Features | Remark |
|---|---|---|---|---|
| 4. | New York State Documents [122] | SS agitated fluidized bed dryers test rig and pilot plant | Test Rig: two samples of residual sludge after anaerobic digested and mechanically dewatered and one more, from outlet of municipal sludge dewatered | • Though the system was not operated at steady state due to supply of the sludge from anaerobic digester calculations and other results presented based on the design specifications and estimated performance.<br>• For a given case huge savings of resources water, thermal energy, electrical energy and reduction in pollutants such as methane, $NO_x$, $CO_x$, etc. Overall 1–2 years payback period was noticed compared to conventional systems. More than 25% energy saving. |
| 5. | Berghel and Renström [130,131] | Industrial size SS fluidized bed superheated dryer (FBSSD) and fixed bed SSD | Biomass: Sawdust and willow wood chips/experimental analysis to verify the control and stability of the dryer | • During drying few design parameters flow rate of feed and superheated steam, temperature, pressure drop, etc. need to be tuned as per requirement.<br>• Even very small steam leaks are devastating to the energy efficiency of the dryer. From experience, the material input and output are very difficult to seal.<br>• The pressure drop over the fixed bed was considerably greater when drying willow wood chips than across fluidized bed when drying sawdust. |

(*Continued*)

**TABLE 4.2 (*Continued*)**
**Details of the SSD with Applications and Features**
**(A) Superheated Steam Fluidized Bed Dryer (SSFBD)**

| S.N. | Paper Details with Authors | Type of Dryer | Feedstock/Type of work/Features | Remark |
|---|---|---|---|---|
| 6. | Taechapairoj [34] | Batch SS fluidized-bed dryer lab scale | Paddy/characteristic study- head rice yield, whiteness and white belly, drying behavior | • Minimum fluidizing velocity for paddy in superheated steam is approximately 2.6 m/s.<br>• Head rice yield from the superheated steam drying is more sustainable and has higher values than those obtained from the hot air drying over a wider range of final moisture contents; however the color of white rice becomes darker.<br>• To maintain good head rice yield, it is recommended that high moist paddy should not be dried to lower than 18% d.b. |
| 7. | Soponronnarit [34] | Continuous SS fluidized-bed dryer lab scale | Parboiled rice/mathematical model validation, characteristic study- pasting viscosities, white belly, hardness, whiteness, and head rice yield. | • Superficial velocity of steam from 1.3 to 1.5 times of the minimum fluidization velocity had no significant effect on the drying rates of rice.<br>• The energy consumption for reducing the moisture content of paddy from 0.43 to 0.22 kg/kg dry basis was approximately 7.2 MJ/kg water evaporated the drying time was approximately 4–5 min.<br>• Head rice yield as compared to that of raw rice is highly improved and obtained higher than 60%. |
| 8. | Wathanyoo et al. [61] | Fluidized bed test rig | Paddy/batch operation head rice yield, whiteness, white belly, viscosity of rice flour and change of microstructure of rice kernel | • For the same drying time, drying rates of SSD paddy were lower than those HAD due to an initial steam condensation<br>• Promoted starch gelatinization improves head rice yield for SSD<br>• Whiteness of SSD paddy was lower than HAD by around 2%. |

*(Continued)*

**TABLE 4.2 (*Continued*)**
**Details of the SSD with Applications and Features**
**(A) Superheated Steam Fluidized Bed Dryer (SSFBD)**

| S.N. | Paper Details with Authors | Type of Dryer | Feedstock/Type of work/Features | Remark |
|---|---|---|---|---|
| 9. | Prachayawarakorn [62] | SS fluidized beds test rig | Soybean/experimental and analytical drying rate, inactivation of antinutritional factors, protein solubility | • For temperature of 135°C–150°C for the HAD and below 135°C for the SSD, the SSD type heating medium shows the protein solubility of treated sample to be higher than HA to the dry soybean.<br>• For the moist soybean, the types of heating medium do not impact on the protein solubility. |
| 10. | Kozanoglu et al. [87,111,124] | Low pressure (vacuum) SS fluidized bed dryer | Coriander, pepper seeds and paddy/ drying kinetics and product quality | • Initial condensation on feedstock is avoided by additional heating of column.<br>• Coriander and pepper seeds showed higher drying rates and lower final moisture contents by increasing operating temperature.<br>• Degree of superheating was the most influential parameter (especially between 2°C and 4°C) than increases of degree of vacuum.<br>• 90°C and 110°C are considered the operating temperatures for superheating steam drying process. |
| 11. | Klutz et al. [132] | SS fluidized bed industrial systems development | Low moisture coal/development of SS fluidized bed dryer at industrial scale | • RWE (Germany based energy company) had been researching the steam drying processes since about 1990 [121] and by now holds 22 patents related to the process.<br>• The construction and operation of the WTA (German abbreviation for 'fluidized-bed drying with internal waste-heat utilization') plants in the various development stages demonstrate that industrial-scale maturity for evaporating capacity of 100 t/h has been reached. |

(*Continued*)

**TABLE 4.2** (*Continued*)
**Details of the SSD with Applications and Features**
**(A) Superheated Steam Fluidized Bed Dryer (SSFBD)**

| S.N. | Paper Details with Authors | Type of Dryer | Feedstock/Type of work/Features | Remark |
|---|---|---|---|---|
| 12. | Shi et al. [133] | SS Fluidized bed test rig | Rapeseed/at normal atm pressure; numerical and experimental verification. | • The temperature and moisture content of particles distributed inhomogeneously in space in the condensing and heating period but distributed uniformly in space in the remaining drying time, and the material was dried evenly without local overheating.<br>• A quality of rapeseed with SSFBD: a good appearance, low shrinkage, and high rehydration ratio than HAFBD.<br>• The minimum fluidized velocity was 1.56*Umf determined by the drying experiment. |
| 13. | Stokie [63] | SS fluidized bed test rig | Victorian brown coals/experimental and modeling/drying rate, moisture readsorption | • Relative drying ratios of SSD and HAD for a similar conditions (such as 130°C–170°C) remains consistent.<br>• For a similar particle size change, the SSD coal has a greater proportional decrease in drying time, and has been attributed to a reduced condensation mass.<br>• Moisture readsorption is lower for SSD than HAD and average difference in it is 1.6%.<br>• Midilli–Kucuk model gives best fit for both SSD and HAD fluidized beds. |
| 14. | Jensen and Larsen [30] | Pressurized SS fluid bed dryers from fundamental research to industrial plants | Sugar beet pulp/Industrial product development-5–100 t/h water evaporation capacity | • One of the currently operating systems (at Idaho, US) evaporating 70 ton/h water, saves 200 tons coal per day and does the drying without air pollution.<br>• The pressurized steam drying technology is mainly interesting when a larger quantity of water will be evaporated.<br>• May used for many other applications as solid waste left after processing like Mash from beer brewing and others (distillers grain), biomass- wood chips, bark, sawdust, bagasse, brown coal, sludge etc. |

**TABLE 4.2 (*Continued*)**
**Details of the SSD with Applications and Features**
**(A) Superheated Steam Fluidized Bed Dryer (SSFBD)**

| S.N. | Paper Details with Authors | Type of Dryer | Feedstock/Type of work/Features | Remark |
|---|---|---|---|---|
| 15. | Aziz et al. [134] | SS fluidized bed test rig | Empty fruit bunch (solid waste from crude palm oil mills)/trial on lab scale model-influence of target moisture content and fluidization velocity | • Up to 92% of the energy involved in the drying process can be recirculated.<br>• The total energy consumption for drying decreases as the target moisture content also decreases, with no significant impact of fluidization velocity.<br>• Required total length of the heat transfer tubes immersed inside the fluidized bed dryer is calculated because it relates to fluidization performance and economic issues.<br>• Lower target moisture content results in a longer heat transfer tube, and higher fluidization velocity leads to a shorter heat transfer tube.<br>• Proposed integration of drying and IGCC seems to be promising technology in terms of energy efficiency. |
| 16. | Jittanit [65] | SS fluidized bed test rig | Parboiled rice/experimental<br>Laboratory-scale Two stage-first stage: SSD and second stage: HA oven drying and FBD/<br>Chalkiness and low milling, gelatinization of starch | • The hot water soaking combined with multi-stage intermittent drying method using SSD provided parboiled rice products with comparable or superior quality.<br>• The superheated-steam drying combines the steaming and drying steps into one process and consumes less energy if the exhaust steam from SSD is recycled. |

**TABLE 4.2**
**Details of the SSD with Applications and Features**
**(B) Superheated Steam Pneumatic Conveying Dryers/Flash Dryer (SS-PCD/FD)**

| S.N. | Paper Details with Authors | Type of Dryer | Feedstock/Type of Work/Features | Remark |
|---|---|---|---|---|
| 1. | Hulkkonen et al. [135] | Pilot scale Pressurized flash dryer | Peat and saw dust and crushed wood chips of pine; No heat exchangers used in the dryer itself. Parameter variation: steam temperatures and velocity. | • Pressures up to 27 bar and temperatures up to 400°C.<br>• Integrated with gas turbine power plant of drying system is feasible and can be commercialized. |
| 2. | Fyhr [136] | SS Pneumatic conveying dryers (Flash dryer) pilot plant | Wood chips/Pilot test- to validate single particle and hydrodynamic model with experimental data<br>U shaped dryer loop | • Single particle model is two-dimensional in order to account for the strong anisotropicity of wood and contains the main mechanisms of heat and mass transport which occur in wood during drying. The hydrodynamic model is one-dimensional and calculates the profiles of temperature, pressure and slip-velocity along the dryer.<br>• The effects of steam properties such as temperature, pressure and steam velocity on drying were investigated. The chip size and wood species also affect drying and were therefore varied. Different design parameters such as height and diameter of the dryer were also varied in order to understand their effects on drying performance. |
| 3. | Blasco and Alvarez [137] | SS pneumatic conveying dryers (flash dryer) pilot plant | Fish meal/experimental and model validation/pilot plant of 60 m length | • Flash drying with superheated steam, under isothermal conditions, is technically feasible.<br>• The Torrezan Santana correlation corrected for mass transfer makes good predictions of the heat transfer coefficient in superheated steam.<br>• For final moisture prediction, variable diffusivity model with the empirical coefficients gives good predictions. |
| 4. | Blasco et al. [138] | SS spouted bed Pneumatic conveying dryers (Flash dryer) pilot plant | Carenite (kyanite)/variation of parameter study and feasibility. | • 2D (bi-dimensional) model, radial and axial is explaining the physics and experimentally validated.<br>• Dense phase transport occur only at gas velocity below 23 m/s with solid flow rates range 0.18–0.37 kg/s. |

**TABLE 4.2**
**Details of the SSD with Applications and Features**
**(C) Superheated Steam Rotary Dryer (SSRD)**

| S.N. | Paper Details with Authors | Type of Dryer | Feedstock/Type of Work/Features | Remark |
|---|---|---|---|---|
| 1. | Chryat et al. [125] | Industry scale SS rotary dryer: cocurrent-triple pass design | Pressed beet pulp/industrial scale demonstration-synchronization of various operating conditions: temperature, pressure, flow rate of feed and steam | • Test performed for pressed beet pulp 27% dry matter at 240°C–300°C, 1.2 bar and received promising results.<br>• There are challenges of feeding and extraction of the material mainly leakages of steam. |

**TABLE 4.2**
**Details of the SSD with Applications and Features**
**(D) Superheated Steam Impingement Drying (SSID)**

| S.N. | Paper Details with Authors | Type of Dryer | Feedstock/Type of Work/Features | Remark |
|---|---|---|---|---|
| 1. | Douglas [139] | Lab scale continuous SS Impingement | Paper/to treat paper properties—drying process relations as well as process engineering aspects | • Paper made from mechanical pulps drying in superheated steam produces a better bonded sheet which is thereby stronger and has a lower scattering coefficient.<br>• Surface properties of such steam dried paper are improved, including reduced linting.<br>• Desorption equilibrium shows that completely dry paper can be obtained at very low superheats.<br>• For superheated steam impingement drying, rates of moisture removal can be about twice as high as that of drying with hot air. |
| 2. | Shiravi [140,141] | Lab scale continuous SS Impingement cylindrical drying plant | Black liquor slurry/design, development, and parametric analysis to check feasibility.<br>Drying kinetics-steam temperature and flow rate, rotor speed | • Drying of black liquor to solids contents 67% to be raised in excess of 92%.<br>• A two-section steam hood. the optimal operating conditions were found to be: SS Jet temperature = 350°C velocity = 6.7 m/s in section I. and temperature = 200°C; velocity = 27 m/s in section II of the hood.<br>• The average drying rate achieved is too low to make the process economical even after improvements made to the process. The limitations of maximum jet velocity and temperature do not provide the potential to increase the drying rate to much higher levels.<br>• This dryer might be useful for supplying either an additional mill evaporation capacity or an additional evaporation level (80%–92% solids). |

(*Continued*)

**TABLE 4.2 (*Continued*)**
**Details of the SSD with Applications and Features**
**(D) Superheated Steam Impingement Drying (SSID)**

| S.N. | Paper Details with Authors | Type of Dryer | Feedstock/Type of Work/Features | Remark |
|---|---|---|---|---|
| 3. | Bonazzi et al. [142] | Review of food drying technology including superheated steam dryers: Cross flow porous belt dryer, Rotating drum dryer, fluidized bed dryers, spray dryer | Chopped spent sugar beet cossettes, liquids and sludges | • Plant developed by Bertin-EDF France similar to multi-stage evaporator plant<br>• Promill, France- rotating drum with pressurized up to 3 bar.<br>• Authors discussed that SSD is till budding technology being dryers are still in development phase and recommended more laboratory studies with diverse products. |
| 4. | Bórquez [143] | Pilot SS impingement cylindrical drying plant | Pine sawdust and mackerel press-cake. Feed rate, residence time of solids, dryer inclination, retention of omega-3 fatty acids | • The transport of solids through the impingement corotational system depends on dryer inclination, gas nature, and gas flow rate. Inclination angle 10°–20° with residence times sufficiently short (113–12 s, respectively).<br>• Drying rates, and so moisture removal and heat and mass transfer coefficients, are increased at higher dryer inclinations and higher gas temperatures.<br>• Drying of mackerel press-cake with superheated steam has low residence times, and has a dried product with high moisture removal with low chemical loss of omega-3 fatty acids and is technically more beneficial than hot air. |
| 5. | Choicharoen et al. [144] | Lab scale coaxial SS impinging dryer | Okara (soy residue)-parametric analysis-volumetric heat transfer coefficient and volumetric water evaporation rate, superheated steam temperature, steam velocity, material feed flow rate and the dryer geometric parameter viz. impinging distance | • Maximum volumetric water evaporation rate around 807 kg water/$m^3$h, while the maximum volumetric heat transfer coefficient around 7950 W/$m^3$K at the steam recycle ratios of 46%–63%.<br>• Lowest total specific energy consumption of the system was around 3.1 MJ/kg water at an inlet steam temperature of 190°C, inlet steam velocity of 20 m/s, material feed flow rate of 20 kg dry solid/h, impinging distance of 5 cm and steam recycle ratio of 63%.<br>• 46% and 95% savings in the total specific energy consumption as compared with the cases of hot air SD system with and without exhaust air recycle, respectively, were achieved. |

*(Continued)*

**TABLE 4.2 (*Continued*)**
**Details of the SSD with Applications and Features**
**(D) Superheated Steam Impingement Drying (SSID)**

| S.N. | Paper Details with Authors | Type of Dryer | Feedstock/Type of Work/Features | Remark |
|---|---|---|---|---|
| 6. | Swasdisevi et al. [145] | Lab scale coaxial SS Impinging stream dryer | Paddy (Phitsanulok 2 variety); inlet air temperatures of 150°C, 170°C, and 190°C; inlet steam velocity of 20 m/s; impinging distance; volumetric heat transfer coefficient and volumetric water evaporation rate | • Two-pass drying, an increase in the drying temperature led to a decrease in the volumetric heat transfer coefficient; however, did not affect the volumetric water evaporation rate.<br>• The SEC for superheated steam as the drying medium was lower than that when using hot air as the drying medium at the same drying temperature.<br>• Superheated-steam drying with 90% recycle could conserve more energy than with 60% recycle.<br>• Color of the dried paddy do not affect by the change in the drying temperature in all cases; superheated-steam drying led to slightly redder and more yellow dried paddy.<br>• The percentage of head rice yield decreased with an increase in the drying temperature; superheated-steam drying led to higher head rice yield than hot air drying when considered at the same drying temperature.<br>• The degree of starch gelatinization was highest when drying was conducted at 150°C. |

**TABLE 4.2**
**Details of the SSD with Applications and Features**
**(E) Superheated Steam Spray Drying (SSSD) at Atmospheric and Vacuum Pressure**

| S.N. | Paper Details with Authors | Type of Dryer | Feedstock/Type of Work/Features | Remark |
|---|---|---|---|---|
| 1. | Dolinsky | Industry scale superheated steam spray drying- high temperature | Materials with thermal sensitivity. Alcohol solutions and suspension, ceramic suspensions, ferreted, various solutions of inorganic salts, etc. | • Temperature of a chamber wall may fluctuate within the range from 200°C to 650°C, and temperature of the superheated steam in the chamber may vary from 100°C to 150°C.<br>• Control of liquid dispersion and temperature regime is also possible. |
| 2. | Islam et al. [146] | Lab scale vacuum superheated steam spray drying | Concentrated orange juice-moisture content, hygroscopicity, water activity, particle size, particle morphology, color characteristics, rehydration and ascorbic acid, maltodextrin concentration, and drying conditions | • Low-temperature (40°C–50°C) and vacuum pressure 5 kPa drying powderization of liquefied food using superheated steam (200°C).<br>• Four different combinations of juice solids: by weight Maltodextrin solids at 60:40, 50:50, 40:60, and 30:70<br>• The dried orange powder is stable due to their low moisture content (2.29%–3.49%), low water activity (0.15–0.25) and higher glass transition temperature.<br>• Particle sizes of orange powder were smaller, smooth and spherical in morphology.<br>• Maltodextrin solids 30:70 are feasible for industry application due to better stability and product recovery. |

*(Continued)*

**TABLE 4.2** (*Continued*)
**Details of the SSD with Applications and Features**
**(E) Superheated Steam Spray Drying (SSSD) at Atmospheric and Vacuum Pressure**

| S.N. | Paper Details with Authors | Type of Dryer | Feedstock/Type of Work/Features | Remark |
|---|---|---|---|---|
| 3. | Lum et al. [59,147–149] | Lab scale counter current superheated steam spray drying | Milk powder from skim and whole milk, mannitol and sodium chloride<br>Parameters: wettability, surface composition, solubility and powder morphology of spray dried skim and full cream milk powder | • The solubility of steam dried milk is relatively similar but slightly lower than that of air. Milk particles were observed to retain an overall more spherical structure instead of collapsing as such in air drying which may contribute to a greater surface area in contact with water, leading to the observed enhanced wettability.<br>• At 180°C, SSD produced significantly fine crystal grains relative to that produced via hot air dryer despite at a slower drying rate for Mannitol.<br>• SSSD led to the formation of a unique hollow hopper like salt crystal structures which was not observed in hot air.<br>• The potential use of SSSD as a useful chemical free media for crystallization control to potentially engineer specific crystal structures will be possible as an alternate existing technique, which utilizing the addition of organic ligands and other additives.<br>• Solubility of the SS-dried milk powders (both, skim milk and full cream milk) only decreased slightly when compared to hot air drying. |

(*Continued*)

**TABLE 4.2 (*Continued*)**
**Details of the SSD with Applications and Features**
**(E) Superheated Steam Spray Drying (SSSD) at Atmospheric and Vacuum Pressure**

| S.N. | Paper Details with Authors | Type of Dryer | Feedstock/Type of Work/Features | Remark |
|---|---|---|---|---|
| 4. | Ma et al. [103] | Lab scale superheated steam spray drying (SSSD) | Instant coffee and sodium copper chlorophyllin (natural green colorant) powders- Feasibility study.<br>Yield, physical properties and morphology study | • Superheated steam spray drying with given test conditions might not be a proper method for producing coffee powder as coffee extract adhered and dried on the wall of the chamber.<br>• Superheated steam spray drying, however, has the potential to produce the color powder with smaller particle size, higher bulk density and more wrinkle surfaces, resulting in the superior solubility. |

(*Continued*)

**TABLE 4.2 (*Continued*)**
**Details of the SSD with Applications and Features**
**(E) Superheated Steam Spray Drying (SSSD) at Atmospheric and Vacuum Pressure**

| S.N. | Paper Details with Authors | Type of Dryer | Feedstock/Type of Work/Features | Remark |
|---|---|---|---|---|
| 5. | Linke et al. [150] | Lab scale superheated steam spray drying (SSSD)-pure superheated steam operation at atmospheric pressure with controlled supply | Dairy product: skim milk powder, lactose, micellar casein concentrate (protein concentration 81.6%) and a lactose (4.7%) and fat (1.6%), whey protein isolate with a protein concentration (>92%) and lactose and fat content (<0.2%),<br>Food product: maltodextrin with a low DE (<3), gum Arabic seyal, soy protein isolate, coffee extract.<br>Drying behavior, physical powder properties, water content and activity, particle size, morphology and color, and product quality | • Special attention to media transfer into and out of the system, as well as to insulation to prevent undesired condensation and to maintain a pure superheated steam atmosphere during drying.<br>• Temperature control of the feed prior to atomization and the pure and oxygen-free superheated steam atmosphere inside the drying chamber together with the immediate cooling of powder particles after discharge from the steam atmosphere.<br>• No products are burned; powders differ noticeably in visual color appearance. Dairy products the difference in color is most pronounced for skim milk powder, followed by micellar casein concentrate and lactose. No difference in color is observed for whey protein isolate.<br>• General Food products, maltodextrin DE (<3) is slightly less bright compared to its industrial reference. Gum Arabic seyal and soy protein isolate are noticeably brighter than their reference samples may be considerably smaller compared to their industrial references. Coffee extract shows a stronger brown coloring.<br>• Dairy powders, most particles show a spherical, to a certain extent wrinkled, structure.<br>• Overall, drying of exemplary dairy and food materials by pure superheated steam has been practically feasible and successfully demonstrated. |

# 5 Low-Pressure SSD

## 5.1 INTRODUCTION

For drying with a superheated steam environment, in addition to the operating temperature of the drying medium, the feedstock temperature is important in the drying process (controlling of the browning, drying kinetics, and other parameters), which is controlled by the drying chamber pressure. Superheated steam dryers can be operated above, at, and below atmospheric pressure, changing the feedstock temperature in a constant drying rate region from very high (more than 100°C) to very low (40°C–80°C) as discussed in the first chapter [9]. When working with a near atmospheric pressure of a superheated steam dryer, the feedstock temperature is approx. 100°C in a constant drying rate period, which may be higher for heat-sensitive substances (certain food products such as vegetables, fruits, herbs, and spices; ceramic components, etc.), causing burning of them and hence the advantages of superheated steam as drying medium are not possible. To have feasible superheated steam drying at a lower temperature range (40°C–80°C), it needs to operate at a lower than atmospheric pressure (vacuum pressure), reducing the temperature of the feedstock less than 100°C in a constant drying rate period (approximately equal to saturation temperature at that pressure), so that advantages of superheated steam drying media may be exploited such as high drying rate, lower drying time, better product quality, etc. This is generally referred as low-pressure superheated steam drying (LPSSD), also referred as vacuum steam drying by some researchers.

### 5.1.1 PRINCIPLE OF LPSSD

The vaporization temperature of water depends upon the pressure acting on it. As the water pressure is lowered below atmospheric pressure making vacuum above its surface appropriately, it will vaporize below 100°C. To adopt this, the vacuum pump is used, which lowers and then maintains the pressure of drying chamber below atmospheric pressure in LPSSD. This will reduce the vaporization temperature of the superheated steam supplied in it accordingly, which controls the temperature of the feedstock, generally maintains at the saturation temperature in constant drying rate region (e.g., for pressure 10 kPa ab, saturation temperature 45.8°C). A line diagram shown in Figure 5.1 explains the principle of the LPSSD.

Generally, 6–8 bar steam generated by steam generator is supplied at saturated condition, which cannot be directly supplied to drying chamber in this case. So there is a way of mounting steam reservoir with pressure regulating valve, maintaining saturated pressure at 1.5–2 bar [26]. From this, it is supplied to drying chamber, which is maintained at vacuum pressure, resulting in the superheated steam immediately as the temperature of the supply steam is sufficiently higher than saturated

DOI: 10.1201/9781003275299-5

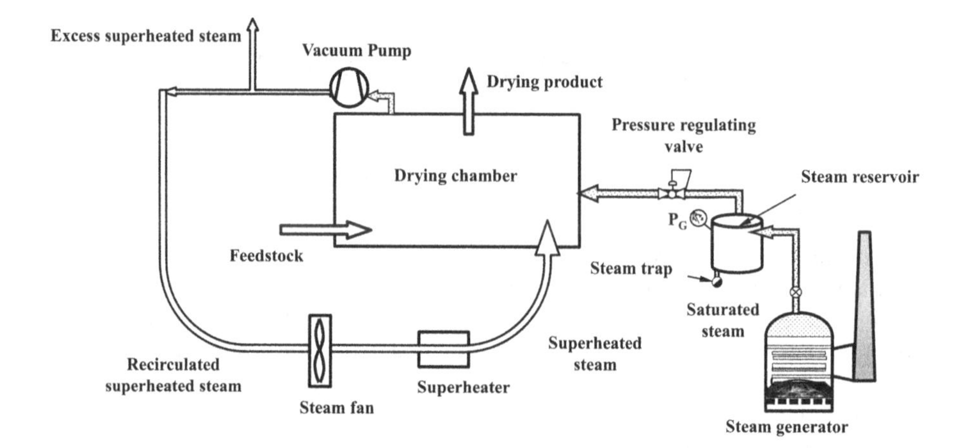


**FIGURE 5.1** A representative sketch of general low-pressure superheated steam drying.

temperature of the steam at dryer pressure (maintained at vacuum). Additionally electric heating coil/thermic oil or high-pressure steam heating system is installed at drying chamber to accurately control the temperature of superheated steam. The exhaust steam is higher than the supply steam by the amount of moisture evaporated from the feedstock. The part of exhaust steam may be recirculated back into drying chamber after preheating to initial conditions, if feasible otherwise its energy content can be recovered. The surplus steam may be needed to use appropriately to have higher performance of the plant.

### 5.1.2 Advantages and Challenges of LPSSD

LPSSD is having all advantages mentioned in Chapter 1 in addition to the following. LPSSD will be applicable for drying of the heat-sensitive materials due to lower saturated temperature of the superheated steam. Further, in general effective inactivation of microorganisms is possible in water activity conditions [151], indicating LPSSD as a more appropriate drying technique than the other techniques for the same. There are a few challenges such as maintaining vacuum is a critical issue and it also increases the operating cost. LPSSD is feasible but not economical for most high-tonnage products. The feedstock insertion and dry product extraction from drying chamber, avoiding infiltration, and maintaining vacuum pressure are other challenges. This technology is not yet matured and commercialized and needs a lot of research.

The important work available in open literature on LPSSD till date is presented in tabular form as shown in Table 5.1A. Most of the work is on laboratory test-rig or lab-scale model and showing positive and motivating results for diverse materials. One of the applications for the orange juice powder with low-pressure spray superheated steam dryer (LPSSSD) [146] is described herewith.

**TABLE 5.1A**
**Details of the Low-Pressure Superheated Steam Dryer (LPSSD)**

| S.N. | Paper Details with Authors | Type of Dryer | Feedstock/Type of Work/Features | Remark |
|---|---|---|---|---|
| 1. | Mujumdar [9] | LPSSD packed bed | Silk cocoon | • For steam drying of silk, higher temperature is possible than hot air drying.<br>• The initial work claims superior quality of silk dried in LPSSD than hot air, however, detailed and exhaustive study need to be performed. |
| 2. | Shibata et al. [161] Mata et al. [162] | LPSSD lab test-rig | Sintered Spheres of Glass Beads/ Experimental and Analytical modeling for prediction of drying rate during constant and falling rate region and critical moisture/Study of drying kinetics and variation of vacuum pressure | • The critical moisture content is not constrained by pressure in low-pressure superheated steam drying (LPSSD) and is lower than hot air vacuum drying.<br>• In falling drying rate region, LPSSD has greater drying rate than hot air vacuum drying. |
| 3. | Devahastin et al. [35] | Lab test-rig LPSSD | Carrot cubes/Experimental and modeling heat-sensitive material as carrot. Effluence of operating parameters-pressure and temperature on the drying characteristics as well as quality attributes, i.e., volume, apparent density, shrinkage, and rehydration behavior, and color/Comparison of LPSSD with vacuum drying | • Drying time for VD is lower by small margin than LPSSD for 7 kPa and 80°C drying conditions for both. Reason may be radiations of electric heaters absorbed by carrot surface in addition to initial condensation.<br>• The quality of dried carrot is superior to VD.<br>• Effect of temperature is more significant than pressure of drying chamber on the drying kinetics.<br>• Over the studied operating parameters, LPSSD delivered enhanced rehydration and a redder dried carrot than vacuum drying. |
| 4. | Suvarnakuta et al. [85] | Lab test-rig LPSSD and comparative study with VD and HAD | Carrot cubes/Experimental and modeling for quality parameters/β-carotene degradation study | • LPSSD is better than VD and HAD having less degradation of β-carotene (@ 20%–25%). |

*(Continued)*

**TABLE 5.1A (*Continued*)**
**Details of the Low-Pressure Superheated Steam Dryer (LPSSD)**

| S.N. | Paper Details with Authors | Type of Dryer | Feedstock/Type of Work/Features | Remark |
|---|---|---|---|---|
| 5. | Kozanoglu et al. [87]<br>Kozanoglu et al. [124] | Lab test-rig Fluidized bed LPSSD<br>Fixed bed of 10 cm of height | Coriander and pepper seeds/Additional heating in fluidized bed avoids pre-condensation/90°C–110°C and pressure-40–66 kPa (ab), drying kinetics.<br>Measurement of bed operating pressure: 1. vacuometer mounted in the downstream of the column of system, 2. differential manometer using mercury located at the top of the column and the atmospheric pressure.<br>Mass flow rates: (0.0049–0.0134 kg/s)<br>Steam velocities: (2.35–4.10 m/s) | • Both (coriander and pepper seeds) particles showed higher drying rates (lower drying time) and reduced final moisture contents with increasing operating temperature.<br>• Temperature of drying media (SS) is having higher influence than pressure lower than atmospheric (vacuum) and the superficial gas velocity on drying kinetics and is strong controlling parameter.<br>• A rotary pump establishes vacuum conditions in the column.<br>• Optimal configuration: 110°C and pressure-40 kPa (ab) of SS in drying chamber. |
| 6. | 2007 Nimmol et al. [163] | Lab test-rig Low-pressure superheated steam and far-infrared radiation drying at the same time (simultaneous) | Banana/lab scale test-rig/Effect of drying medium temperature and pressure on the drying kinetics, heat transfer, energy consumption; Comparison with combined far-infrared radiation-vacuum drying (FIR–VD) and LPSSD; | • The drying rate reduces with increasing temperature and reducing drying chamber pressure due to increasing driving potential.<br>• On the basis of drying rate and specific energy consumption at all tested configurations, LPSSD–FIR with 90°C and pressure 7 kPa (ab) was most optimal.<br>• LPSSD gives higher drying time at 90°C and 7 kPa (ab) than LPSSD-FIR and VD-FIR.<br>• In addition to drying time specific energy consumption, quality parameters need to be studied in detail. |

(*Continued*)

**TABLE 5.1A (*Continued*)**
**Details of the Low-Pressure Superheated Steam Dryer (LPSSD)**

| S.N. | Paper Details with Authors | Type of Dryer | Feedstock/Type of Work/Features | Remark |
|---|---|---|---|---|
| 7. | 2007 Pimpaporn et al. [82] | Lab test-rig LPSSD, Various combined pretreatments<br>A liquid ring vacuum pump | Potato chips/Drying kinetics, microstructure, and quality parameters (colors, texture- toughness, hardness, and crispness) 70°C, 80°C, and 90°C, pressure 7 kPa (ab) | • To minimize pre-condensation during the start-up period in the drying chamber, special heater is provided.<br>• LPSSD at 90°C along with pretreatments (combined blanching and freezing) was suggested as the most conducive conditions for potato chips drying.<br>• It is having smaller drying time, better product quality, and less thermal damage at above condition than others. Sensory study is suggested. |
| 8. | 2007 Thomkapanich et al. [164] | Lab test-rig Low-pressure superheated steam | Banana/Drying kinetics, behavior heat transfer of the banana chips/Quality of product: texture, shrinkage, retention of ascorbic acid, color, consumption of Energy and saving due to intermittent SS and vacuum compared to conventional hot air and vacuum drying.<br>Initial moisture content 0.025 kg/kg (d.b.) | • Intermittent steam supply for on:off period 10:20 and 90°C LPSSD reduced net drying time resulting in 65% savings in energy compared to continuous LPSSD, but, no effect noticed on drying rate.<br>• Drying rate for intermittent pressure LPSSD and vacuum drying at all conditions are higher than continuous drying saving in vacuum pump energy by 51% and 53% respectively of LPSSD and vacuum drying. Savings of steam in this case is 58% compared to continuous.<br>• Shrinkage was not affected by intermittent steam but intermittent pressure is having significant effect compared to continuous.<br>• Intermittent steam supply led to higher retention of ascorbic acid especially for longer tempering (off) periods, but intermittent pressure shows greater degradation.<br><br>Overall, there is no much beneficial in case of quality, except retention of ascorbic acid retention and as energy savings. It may be checked for product to product but there is no trend seen about its beneficial effects. |

(*Continued*)

**TABLE 5.1A (*Continued*)**
**Details of the Low-Pressure Superheated Steam Dryer (LPSSD)**

| S.N. | Paper Details with Authors | Type of Dryer | Feedstock/Type of Work/Features | Remark |
|---|---|---|---|---|
| 9. | Kingcam et al. [98] | Lab test-rig LPSSD, Various combined pretreatments<br>A liquid ring vacuum pump (Same of Pimpaporn et al. [82]) | Potato chips/initial slice thickness, degree of starch retrogradation, and final moisture content, pretreatment methods, Potato slices thicknesses 1.5, 2.5 and 3.5 mm | • Greater intensity of starch retrogradation led to a rise in the toughness and hardness with a higher rate of crystallinity of dried chips, while no effect on the crispiness. Full sensory assessment is recommended. |
| 10. | Tang [165] | Lab test-rig LPSSD | Distiller's spent grain (DSG)/ Experimental/Parametric analysis-temperature: 95°C–115°C and pressure: 76 or 81 kPa (ab), SS velocity: 0.1–0.289 m/s | • Engineering issues of low-pressure (vacuum) in drying chamber, infiltration due to leakages, and sudden temperature drop were resolved and LPSSD was successfully used to dry DSG.<br>• Parametric analysis also performed by varying the SS temperature and velocity, and chamber pressure. |
| 11. | Kozanoglu et al. [111] | FBLPSSD lab scale test setup | Paddy/Experimental Drying kinetics of paddy/40–67 kPa, 98°C–118°C, superficial steam velocities (2.9–4.0 m/s), mass flow rates (0.0061–0.0103 kg/s) | • Additional heat supplied into the column by electrical resistances eliminates the pre-condensation.<br>• Drying rates increase by increasing operating temperature, however, operating pressure and superficial steam velocity have limited effect over drying process. Further, there is hardly any effect in the falling drying rate period observed.<br>• Higher degrees of superheating generate lower equilibrium moisture contents.<br>• LPSSD for fluidized bed is feasible to perform at lower temperature. |

(*Continued*)

**TABLE 5.1A (*Continued*)**
**Details of the Low-Pressure Superheated Steam Dryer (LPSSD)**

| S.N. | Paper Details with Authors | Type of Dryer | Feedstock/Type of Work/Features | Remark |
|---|---|---|---|---|
| 12. | Devahastin and Mujumdar [26] | Review of LPSSD and case study on Carrot drying in lab-scale LPSSD | LPSSD from various studies in the lab of Prof. Devahastin/Book chapter | • LPSSD is important for heat-sensitive food and biomaterials and need to explore its full potential.<br>• The qualities of LPSSD materials are superior. |
| 13. | Messai et al. [166,167] | Lab test-rig LPSSD | Porous media spherical particles: ceramic, coal, etc./Modeling for drying characteristics of single particle; determination of inversion temperature by theoretical model. | • Proposed model based on average volume theory to calculate inversion temperature for LPSSD.<br>• Inversion temperature for constant and falling rate periods are different as per the model and it is also depends on drying chamber pressure, particle diameter, and particle permeability.<br>• Increase in drying chamber pressure results in the reduction in the moisture mass flow, and a rise in the temperature of drying media results in an increase in moisture mass flow during both, constant and falling rate periods.<br>• Reduction in chamber pressure and increase in temperature results in an increase in moisture mass flow for the product with high porosity. |

(*Continued*)

**TABLE 5.1A (*Continued*)**
**Details of the Low-Pressure Superheated Steam Dryer (LPSSD)**

| S.N. | Paper Details with Authors | Type of Dryer | Feedstock/Type of Work/Features | Remark |
|---|---|---|---|---|
| 14. | Islam et al. [146] | Low Pressure Spray Superheated Steam Dryer<br>Upward flow and atomized by dry hot air.<br>Maintained the supply of superheated steam by separate nozzles appropriately. | Orange juice powder- four combinations of solids in orange juice : maltodextrin for 60:40, 50:50, 40:60, and 30:70/<br>Low-temperature (40°C–50°C)<br>Physicochemical properties-<br>Moisture content, water activity, hygroscopicity, particle size and morphology, color characteristics, rehydration and ascorbic acid retention.<br>Water sorption and glass transition temperature of juice powder of orange conditioned at various water activities | • Heating medium: Superheated steam (200°C) in the spray dryer chamber and cyclone separator.<br>• Maltodextrin concentration and drying conditions have a significant influence on the properties of the dried powder.<br>• Stable orange powder due to smaller moisture content (2.3%–3.5%) and water activity (0.15–0.25), and greater glass transition temperature, also retained a significant amount of ascorbic acid (71%).<br>• Powder morphology: small, smooth and spherical particles<br>• Optimal value of ratio of solid contents of concentrated orange juice (COJ)/Maltodextrin (MD) 30:70 by weight is useful for commercial production of orange juice powder. |
| 15. | Liu et al. [96] | Lab test-rig LPSSD test-rig | White radish discs/Drying kinetics and quality attributes/95 mbar (ab) and drying temperature 75°C–90°C. | • Rehydration capability higher than vacuum drying (VD).<br>• Around 25% of the total amount of Vitamin C was preserved; further Vitamin C from exhaust steam can be recovered and preserved appropriately from the condensate. |
| 16. | Sehrawat and Nema [94] | LPSSD lab-scale system | Onion slices/Experimental and model fitting for moisture removal/Quality parameters: retention of color, rehydration ratio, thiosulfinate content, total phenol content, and antioxidant activity | • Temperature of drying media is having high influence. Optimal condition for better quality parameters (such as high pungency, better color, and rehydration of dried onion) is at 70°C at 10 kPa (ab). |

(*Continued*)

**TABLE 5.1A (*Continued*)**
**Details of the Low-Pressure Superheated Steam Dryer (LPSSD)**

| S.N. | Paper Details with Authors | Type of Dryer | Feedstock/Type of Work/Features | Remark |
|---|---|---|---|---|
| 17. | Namsanguan and Mangmool [96] | LPSSD at lab-scale test-rig | Longan without stone/Effect of drying temperature and pressure on the drying kinetics and dried product quality viz. color, shrinkage, rehydration and texture (toughness)/70°C–90°C and 7–15 kPa ab | • Initial moisture content in the range of 350%–400% dry-basis was dried until its moisture content reached 18% dry-basis.<br>• Drying rate increased as drying temperature increased and as pressure decreased.<br>• Total color difference seemed to decrease whereas shrinkage increased when drying temperature decreased and pressure increased.<br>• Rehydration and toughness increased as drying temperature and pressure increased.<br>• Considering drying time and product quality, optimal drying condition for SS is 80°C and 15 kPa (ab) for drying of longan without stone. |
| 18. | Malaikritsanachalee et al. [89] | LPSSD and Intermittent LPSSD lab-scale test-rig | Ripe Mangoes/Experimental in lab-scale unit and modeling/Parameters: drying kinetics- drying behavior, drying rate, and effective moisture diffusivity ($D_{eff}$); qualitative properties- color, shrinkage, rehydration, and microstructure. Comparison of LPSSD, Intermittent LPSSD, HAD. | • Drying time of LPSSD was shorter by 58% as compared to the HAD with better quality such as less color changes ($p < 0.05$), shrinkage ($p < 0.05$), and rehydration time than that of the HAD-dried products. Total color differences ($\Delta E^*$) and shrinkage of LPSSD-dried products were lower than HAD-dried products approximately 12.49% and 15.10%, respectively.<br>• The intermittent LPSSD-dried products at 20:1 min provided the highest porous structure and rehydration rate than continuous LPSSD, however there was no significant difference in kinetic, $\Delta E^*$, and S ($p > 0.05$) in all conditions of LPSSD. |

*(Continued)*

**TABLE 5.1A (*Continued*)**
**Details of the Low-Pressure Superheated Steam Dryer (LPSSD)**

| S.N. | Paper Details with Authors | Type of Dryer | Feedstock/Type of Work/Features | Remark |
|---|---|---|---|---|
| 19. | Narmatha et al. [168] | Batch LPSSD at lab test-rig | Potato slices/temperature 70°C–90°C and absolute pressure 0.5–0.8 bar with steam flow rate 0.1–0.3 $m^3$/h/quality parameters-volume, porosity (Ɛ), color, rehydration ratio and percentage shrinkage | • Potato slices dried at 90°C, 0.5 bar (ab) was found to have an enhanced drying rate of 0.1–1.5 g/min with reduced drying time 130–150 min.<br>• Quality attributes are preferred at a lower temperature (70°C) and low pressure of 0.5 bar.<br>• Dried potato slices with a mild cooked flavor and unchanged shape, better texture and color can be obtained by this method.<br>• The steam flow rate did not show any significant effect on both drying characteristics as well as physical attributes |

**TABLE 5.1B**
**Details of the Low-Pressure Superheated Steam Dryer (LPSSD) Applied in Wood Drying**

| SN | Paper Details with Authors | Type of Dryer | Feedstock/Type of Work/ Features | Remark |
|---|---|---|---|---|
| 1. | Pang and Dakin [37] | LPSSD pilot scale plant (Moldrup kiln) | Radiata pine sapwood/drying rate and temperature of Stack for drying characterization, defects such as brown stains and drying stresses/Temperature: 90°C, 0.2 bar (ab) and 10 m/s steam velocity; 125%–160% initial moisture content, 421–472 kg/m$^3$ densities. After drying 12% average moisture content but higher standard deviation 8.3%. | • SSD results in fast drying than high temperature moist air (conventional drying).<br>• Moisture distribution at the end of the drying is variable in the stack.<br>• Defect in drying such as brown stain in kiln and stresses due to drying are reduced due to lower temperature of the stack than conventional drying.<br>• Variation in moisture of stack can be mitigated by flow reversal and high steam velocity as wood near inlet is almost achieved equilibrium moisture (6%) but at the outlet side till wet with 20% moisture.<br>• Overall, better quality (reduced built-up stresses and browning stains) with lower energy consumption and drying time than conventional drying (hot moist air). |
| 2. | Pang and Pearson [169] | Pressurized and low-pressure SS Kiln pilot scale commercial system (Moldrup kiln) | Radiata pine sapwood timber-100 × 40 × 3000 and 1500 mm length with stack 3.0 m long, 1.0 m wide and 0.51 m high. 10 m/s steam velocity<br>Vacuum drying at 90°C/0.2 bar; Pressure drying at 160°C/3 bar<br>Steam temperature and circulation velocity, operation pressure | • Compared to conventional moist hot air drying, LPSS kiln have lower heat capacity due to reduced density having potential benefits of reduction in drying defects such as kiln brown stain and drying stresses, along with reduced energy consumption.<br>• Disadvantage is higher velocity of steam circulation resulting in formation of bypass flow through and around the stack.<br>• High-pressure SSD have five-fold higher drying rate compared to widely use conventional temperature drying with hot moist air.<br>• Ultra high temperature drying with superheated steam |

*(Continued)*

**TABLE 5.1B** (*Continued*)
**Details of the Low-Pressure Superheated Steam Dryer (LPSSD) Applied in Wood Drying**

| SN | Paper Details with Authors | Type of Dryer | Feedstock/Type of Work/Features | Remark |
|---|---|---|---|---|
| 3. | Yamsaengsung and Sattho [170] | Lab scale intermittent SSD-LPSSD drying setup of elliptical vessel | Rubberwood/experimental at lab scale/effect of temperature change on Drying time and mechanical properties: hardness, compression, shear, and bending of the rubberwood. Intermittent SSD-LPSSD schedule: SSD supplied as initial 1 h, then after 4 and 8 h of LPSSD, for 1 h duration | • Drying time reduced by 85% than conventional hot air drying (approx. 7 days) for 25.4 mm rubberwood due to Intermittent SSD-LPSSD (less than 20 h).<br>• From LPSSD to fully SSD, saturated steam takes 10–15 min to superheated steam condition, during this time added moisture helped to relieve build-up stress due to drying and to keep cell pores open eliminating surface cracking.<br>• Optimized condition in this case is 70°C and 12.03 kPa (ab) pressure, where mechanical properties excellent. As temperature increases further, mechanical properties start deteriorating, indicating heat-sensitive nature of rubberwood.<br>• LPSSD-dried rubberwood shows 32% increase in hardness, a 12% increase in compression parallel-to-grain, and 88% increase in shear parallel-to-grain.<br>• More detailed study with pilot testing needed to confirm the test results for variety of thickness and size of rubberwood and for scale-up. |
| 4. | Redman [171] | LPSSD- a 2 m³ capacity vacuum kiln, 0.1–0.2 bar and 40°C–80°C | Four types of hardwood: *Corymbia citriodora* (spotted gum), blackbutt (*Eucalyptus pilularis*), messmate (*Eucalyptus obliqua*) and jarrah (*Eucalyptus marginata*)/parameters: drying quality, time and cost. Develop an economic model and a predictive vacuum drying | • Vacuum drying produces material of the same or better quality than is currently being produced by conventional methods within 41%–66% of the drying time, depending on the species.<br>• Economic analysis indicates positive or negative results depending on the species and the size of drying operation.<br>• Economic benefits exist by vacuum drying over conventional drying for all operation sizes, in terms of drying quality, time and economic viability, for E. marginata and E. pilularis. The same applies for vacuum drying C. citriodora and E. obliqua in larger drying operations (kiln capacity 50 m³ or above).<br>• Overall, LPSSD (vacuum drying) is feasible drying option by all means. |

(*Continued*)

**TABLE 5.1B (*Continued*)**

**Details of the Low-Pressure Superheated Steam Dryer (LPSSD) Applied in Wood Drying**

| SN | Paper Details with Authors | Type of Dryer | Feedstock/Type of Work/Features | Remark |
|---|---|---|---|---|
| 5. | Elustondo et al. [172] | Laboratory-scale superheated steam vacuum drier (Kiln) | Western red cedar, parameters for inspection- grader's personal perception of drying degrade, and the lumber moisture content at the end of the drying | • The drying time for 50 mm western red cedar was under three days with superior quality than that of conventionally dried with significantly lower drying time, which takes two weeks in conventional kilns.<br>• MC after drying is in the range of 10%–15% that is typically tolerated in industry.<br>• Optimal wet-bulb depression observed for given conditions was 2.2°C–3.4°C and small change in this resulted in significant degradation in the quality. |
| 6. | Espinoza and Bond [160] | Review paper: LPSSD for wood drying applications | Various types of wood from research papers. Beech, spruce, and Scots pine, Masson pine, Rubberwood, radiate pine sapwood, birch lumber, Plantation eucalyptus | • Experimental and modeling work from 1999 to 2010 related to wood drying using LPSSD is reported.<br>• Most of the woods are dried in significantly less time with higher drying rate and better product quality.<br>• No direct comparison with other vacuum drying technique reported in it. |
| 7. | He et al. [173] | Lab scale experimental setup | Poplar sapwood, 150 × 100 × 60 mm size. Temperatures of the kiln 35, 55, and 70°C and absolute pressure levels 0.03, 0.06, and 0.1 MPa | • Convective heat transfer coefficients rises with absolute pressure increases at set temperature conditions and rises alongside temperature at set absolute pressure conditions. |

## Case Study 5.1: Low-Pressure Spray Superheated Steam Dryer (LPSSSD) by Islam et al. [146]

The low-temperature superheated steam drying was applied for spray drying of concentrated orange juice with the objective of finding the optimal configuration of maltodextrin and glass transition temperature by Islam et al. [146]. The experiment is performed in a lab-scale spray dryer where a mixture of concentrated orange juice (COJ) and maltodextrin (MD) is atomized through the upward nozzles places at the lower part of the chamber and is pumped at 300 ml/h. Though it uses superheated steam as drying media for atomization, compressed air at 40 l/min (normal pressure and temperature) with pore diameters of 10–50 μm is provided. Superheated steam at 200°C generated by an electric heater is supplied by a separate nozzle in the drying chamber, providing the heat for dewatering. To maintain the feed temperature above the saturation temperature (40°C at 5 kPa), hot water-1 is supplied to maintain the evaporator jacket at 50°C. The dried powder is flowing with superheated steam and is taken out from the spray dryer chamber and passed to cyclone separators 1 and 2. Former captures a significant part of the total powder and from it the powder is taken out appropriately. To avert the development of deposits on the wall and classifications in the cyclone, hot water-2 is supplied to the water jacket of the first cyclone. Further, low-pressure dry air is provided to collect the powder in a receiver at a temperature of 45°C along with hot water-3 in the bottom, which prevents the receiver wall deposition. After use in it, the returned hot water-3 is accumulated at the top of the first cyclone. Similar to the first, the second cyclone is also provided with all auxiliaries to avert the formation of wall deposits and classifications. The superheated steam at the end is condensed by cooling water in the condenser, which also helps to make the vacuum up to a certain level, reducing the size and capacity of a vacuum pump. Figure 5.2 shows the sketch of a superheated steam spray dryer with various components and auxiliaries.

### FEED AND DRYING CONDITIONS

COJ is having various constituents such as total soluble solid (62% ± 0.45%), ascorbic acid (320 mg/100 g), and citric acid (4.8%–5.7%). To avoid stickiness, maltodextrin powder (moisture content 4.15% ± 0.02%) is used as a carrier agent. In this work, total soluble solid (TSS) contents of COJ and maltodextrin solids in various combinations such as 60:40, 50:50, 40:60, and 30:70 (COJ:MD by weight) were chosen for SSSD. The solution combining COJ and MD of 1000 g was prepared having 33 Brix%. The prepared solutions were used for the drying process as feed and the output product as a dried powder, which is evaluated for various physicochemical and functional properties pertinently such as water activity, moisture content, hygroscopicity, particle size and color, morphology, ascorbic acid retention and, rehydration as well as water sorption and the temperature of glass transition.

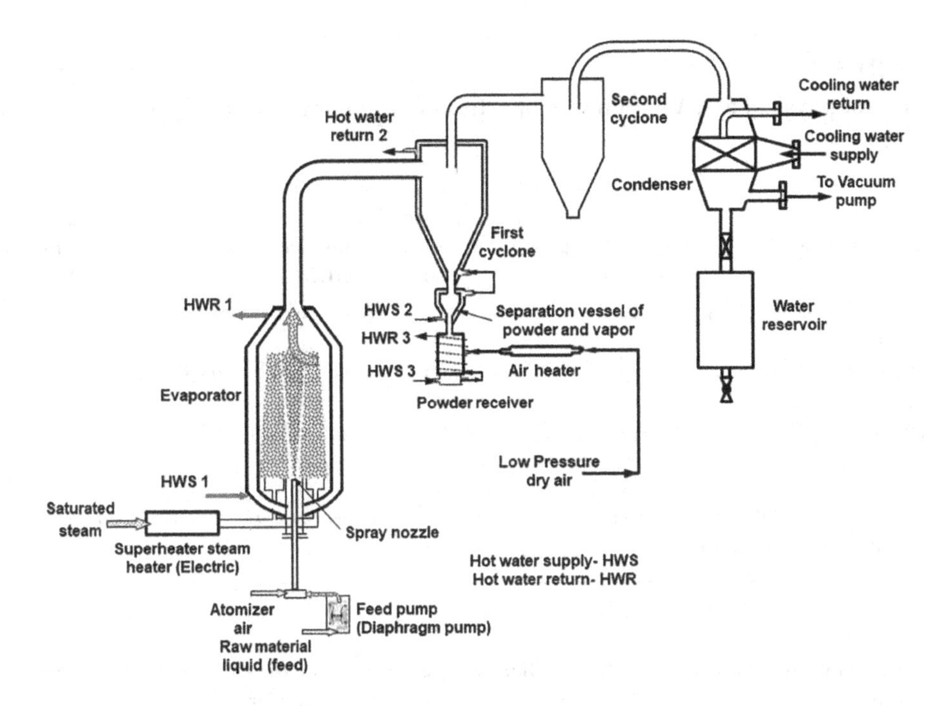


**FIGURE 5.2** Representation of the superheated steam spray drying with process flow [146].

## RESULTS AND DISCUSSION

### Moisture Content, Water Activity, and Glass Transition Temperature for Dried Orange Powder

The orange powder dried by SSSD compared hot air-operated spray dryer [152,153] is having lower water activity ($a_w$ = 0.15 ± 0.01–0.25) and moisture contents (2.29%–3.35%) as proposed in the work of the Islam et al. [146]. In low-pressure superheated steam drying, the temperature gradient for COJ droplets is higher with the superheated steam than hot-air environment due to higher saturation temperature resulting in greater heat transfer, which is the moisture from the droplet at a higher rate. As maltodextrin concentration increases 50% and above, water activity ($a_w$) and moisture content of the dried powder decrease and issues of stickiness also get reduced due to increased glass transition temperature. This is mainly due to the hygroscopic nature of both constituents of the COJ as sugar and acid. Further, for all combinations, water activity for dried powder is below 0.3, which develops the stability of the powder.

It is noticed that with increasing the glass transition temperature ($T_g$) of the dehydrated powder (mixture of CJO and MD), the maltodextrin solid increases because of its higher molecular weight, which is supported by various researchers in literature [154,155]. Further, its increasing percentage will also increase cyclone recovery due to reduced stickiness, and it is maximum for 30:70 CJO/MD, equal to 63.3%, as shown in Table 5.2, which summarizes the experimental results for various combinations of MD. The hygroscopicity and solubility of the dried orange powder increase with

**TABLE 5.2**
**Thermophysical and Vitamin C Properties of Dried Powder of Orange by SSSD [146]**

| Types of Powder | Moisture (%) | Water Activity | Glass Transition Temperature $T_g$ (°C) | Cyclone Recovery, (%) | Hygroscopicity ($gH_2O/g$) | Rehydration (s) | Vitamin C Retention (%) |
|---|---|---|---|---|---|---|---|
| COJ/MD 50:50 | 3.35 ± 0.28[a] | 0.25 ± 0.00[a] | 61.78 ± 0.97[c] | 53 | 0.195 ± 0.02[a] | 253.65 ± 1.05[a] | 71.01 |
| COJ/MD 40:60 | 2.57 ± 0.12[b] | 0.17 ± 0.02[b] | 75.94 ± 0.92[b] | 60.5 | 0.188 ± 0.03[b] | 237.40 ± 1.25[b] | 67.75 |
| COJ/MD 30:70 | 2.29 ± 0.16b[c] | 0.15 ± 0.01[b] | 86.63 ± 0.97[a] | 63.3 | 0.143 ± 0.01[c] | 122.34 ± 1.56[c] | 64.88 |

a, b, c are ± standard deviation about a mean value of the parameter. The means with differnet superscripts in a column differs significantly

increasing the MD percentage as it has excellent solubility in water and rapid rehydration may be due to the lower moisture content of the powder. These observations have also been supported in the literature [152]. It is worth to mention that retention of vitamin C and its content is higher in the lower share of maltodextrin, which is practically possible as 50:50 (CJO:MD), as shown in Table 5.2. This higher percentage may be due to low-temperature operations (low-pressure, 5 kPa ab) and a less-oxygen environment due to the presence of superheated steam drying media. The color of dried powder also has significant effect on MD content as lower the quantity better will be color retention capability. In this work, it is having bright yellow for mixture of 50:50, CJO/MD and reduces it with increase of MD percentages in the CJO/MD mixture as MD is white.

## Particle Morphology of COJ Powder

The field emission scanning electron microscopy (FE-SEM) images show the morphology of the particles of dried powder as shown in Figure 5.3. The powder with CJO/MD (30:70) contents higher maltodextrin had smooth surface, smaller spherical shapes, and no shrinkages (see Figure 5.3a). Powder with lower content of MD than above mixture (40:60, COJ/MD) having smoother spherical shapes, on the other hand with larger particle size than earlier one (30:70, CJO/MD) due to agglomeration of small particles form large one (see Figure 5.3b), and for 50:50 COJ/MD configuration, powder particles are with dented surfaces along with wrinkles and deformation (see Figure 5.3c).

Overall, considering the stability and product recovery, it is noteworthy to put that the orange juice powder with COJ/MD = 30:70 by weight is useful for commercial applications to produce dry powder.

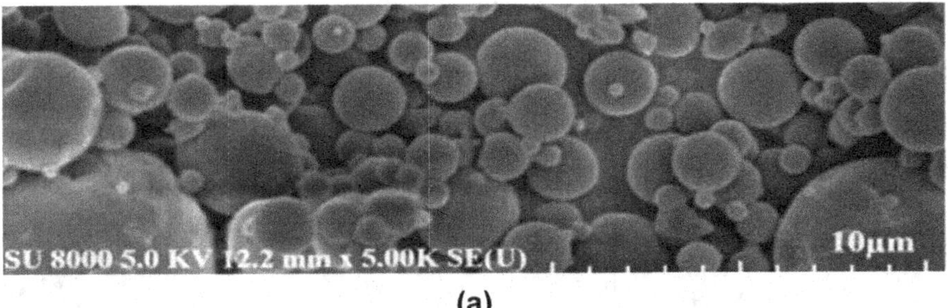


(a)

(b)

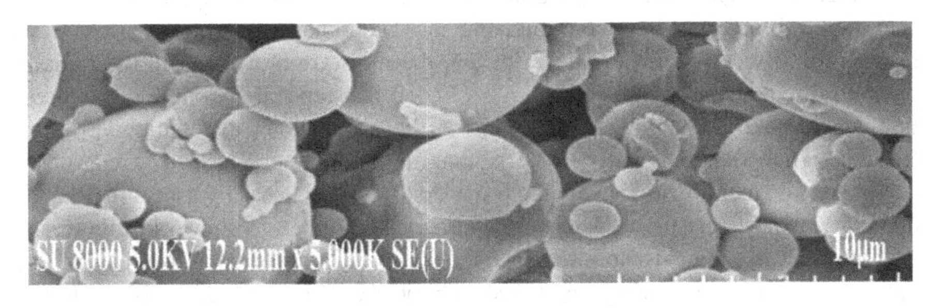


(c)

**FIGURE 5.3** Micrographs of SSSD orange juice powder particles at different COJ/MD ratios took by FE-SEM. (a) 30:70, (b) 40:60, and (c) 50:50 [146].

## 5.2 LPSSD OF WOOD DRYING

Wood is a natural material, mainly made up of organic substances such as cellulose, hemicellulose, and lignin. These organic substances are in the form of cell of cylindrical shape arranged parallelly to each other. Along the longitudinal axis of the wood stem, there are connecting channels for water and minerals transportation from roots to leaves, through which later on moisture gets migrated during drying. Wood after drying has huge applications from ancient times, such as fuel, coal, construction of house, furniture, vehicle, boat, and so on, paper making, engineered

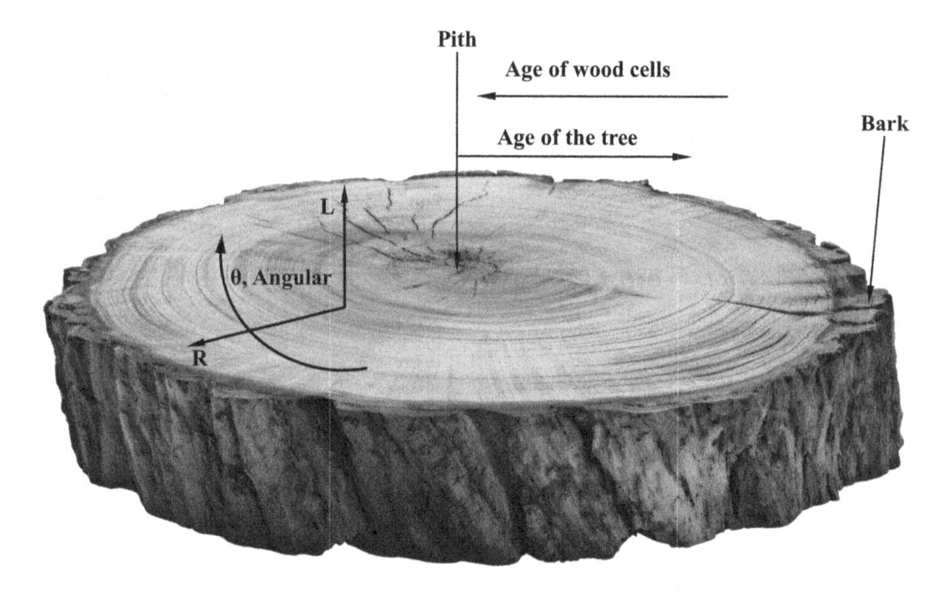


**FIGURE 5.4** Cross-sectional view of the stem of tree. L, longitudinal; R, radial; $\theta$, angular.

wood, wood-derived products like cellulose, cellophane, cellulose acetate, and nanocellulose. However, drying of wood is a complex, time-consuming, and critical process due to limitations on drying rate, and each wood has different microstructures resulting in diverse characteristics. Further, the moisture movement in the radially outward direction across the cell is having higher resistance than longitudinal direction, where natural channels already exist in those directions, as mentioned before. Even though diffusion of water along the longitudinal direction is 10–15 times faster (for the reason mentioned above) than transverse, and this is hardly useful in practice due to the extensively higher length to width of the lumber, as shown in Figure 5.4. Longitudinal diffusion is noteworthy only for the drying of short pieces and the ends of longer boards, having a major contribution to moisture transfer. More details about the plants' internal structure and their effect in drying, along with physics are covered in the book titled *Drying of Aromatic Plant Material for Natural Perfumes* in Chapters 2 and 5 [156].

Generally, green logs drying up to equilibrium moisture content requires 70% of the total energy required for processing the wood. Furthermore, around 10% rejection in quality inspection occurs in dried wood because of surface cracks, distortion, collapse, and moisture variation in addition to these other challenges are the very high drying period, mainly in a number of days and lower drying rate. Due to these critical challenges, more suitable drying techniques are tried over the years, such as superheated steam drying at atmospheric pressure [135,157158] and low-pressure (vacuum).

Generally, in drying of wood, moisture removal mechanisms are in four ways as: (i) movement of liquid water by cell structure through capillary action (free water bulk flow); (ii) transportation of water vapors along the low-pressure zones from high-pressure (water vapor bulk flow); (iii) diffusion of moisture in vapor form because of relative humidity gradients; and, lastly, (iv) diffusion of water molecules from cell walls due to moisture content differences [159]. Fiber saturation point (FSP) is the important stage in the dehydrating or wetting of wood, where the cell walls are saturated with water (bound water) and no water in the cell cavities. Above this (FSP), a controlling parameter for moisture removal is energy transfer, and below it is mass transfer. In wood especially, most of the moisture removal is due to diffusion transfer, where the temperature requirement is significantly higher to maintain the optimal drying rate. The important advantages of vacuum drying of wood over and above commonly described in Section 5.1.2 are (i) reduced warping; (ii) less possibilities of the drying defects such as checking and honeycomb; and (iii) retaining its original color due to lower temperature and oxygen-free environment.

Espinoza and Bond [160] reviewed the vacuum dryers along with low-pressure superheated steam dryers for wood drying application. Latest work in wood drying is appropriately cited in Table 5.1A–C. Overall, wood drying with superheated steam has great potential due to reduced energy consumptions and most importantly significant reduction in drying time. More scientific studies on laboratory as well as pilot scale are needed to explore and tap its full potential.

## 5.3 CLOSURE

The LPSSD is only dryer applicable for heat-sensitive substances. Though the vacuum drying with superheated steam (referred as LPSSD in this book) known from early 1900 its systematic research in large scale started in mid-1980. One of the co-authors Prof. Mujumdar was in leading role to promote it for various diverse applications from paper to silk. Primary advantage of it is lower than 100°C temperature of the feedstock can be possible in constant drying rate region, hence drying temperature can be designed below 100°C as per the requirements. Due to this, heat-sensitive substances can be dried using the superheated steam environment taking the advantages of the steam properties. Currently, good numbers of low-pressure superheated steam dryers are commercially available for various applications such as wood drying, for few vegetables and fruits, spices, coal, ceramics, etc.

In this chapter, out of various types of dryers, mainly spray dryer and fluidized bed at lab-scale and kiln at pilot-scale are discussed. In large commercial systems, one of the critical issue of maintaining the vacuum may be faced. Most of the studies still are at lab test-rig, so to go ahead more pilot-scale studies in diversified applications are needed.

# 6 Integration of SSD with Other Drying Technology

## 6.1 INTRODUCTION

Superheated steam dryers can be easily integrated with other types of dryers as well as other processes to improve overall performance (energy efficiency, drying time, etc.) and or product quality [9]. There are few works on the integration of SSD with other types of dryers, generally referred to as hybrid drying, such as HAD, far-infrared radiation dryers, heat pumps, etc. Furthermore, the excess steam generated during drying, equivalent to the amount of moisture evaporated, needs to be utilized appropriately to improve its green footprint and overall efficiency. This can be best possible by integrating with other processes utilizing the excess steam at 100°C and atmospheric pressure or in thermal power plants, renewable energy sources, etc. Further, there may be a possibility of excess steam compression for energy supply for drying, wherever economically feasible. In this chapter, a discussion on the hybrid drying is put forward, followed by the integration of the dryers in various processes.

## 6.2 HYBRID DRYING PROCESSES

It is a common practice to combine the best features of the different types of the dryer, as optimally combining various drying process will provide several benefits. These are listed as better product quality, lower energy consumption and emission due to exhaust gases, reduced drying time, ease of processes due combining pretreatment and drying process, etc. In this section, hybrid combination of superheated steam dryer and other types of the dryers are presented in Table 6.1 and few of them are explained in more details.

In this domain, mostly lab-scale research work is going on, as it is at the development stage. As discussed in Chapter 1, the working of SSD at lower temperatures requires the operation of the system at vacuum pressure, and maintaining that condition is technically a challenge. Further, if, throughout the drying process, the system works at vacuum pressure, it may take significant time for drying certain products. Therefore, the best approach is to combine the benefits of two different drying technologies and use them in sequence or simultaneously so that the combined system will be an optimal process design.

Nimmol et al. [174] dried bananas using a combination of low-pressure superheated steam drying (LPSSD) and far-infrared radiation drying (FIRD), combined vacuum and far-infrared radiation drying (Vacuum–FIRD), and only


DOI: 10.1201/9781003275299-6

**TABLE 6.1**
**Details of the SSD with Applications and Features**

| S.N. | Paper Details with Authors | Study | Feedstock/Type of work/Features | Remark |
|---|---|---|---|---|
| 1. | Mujumdar [9] | Report on development and stage of commercialization of various SSD and application of materials | Review report on aspects of use of excess steam for energy recovery or other plant application, recompression of the vapors, various types of the compressors and appropriate compressor for steam. | • Mechanical vapor compression technology is interesting but steam jet ejectors (thermocompression) may be cost effective alternative to recompress exhaust steam for recycle.<br>• New technologies in steam compressors are important for some major drying applications.<br>• Recirculation of the exhaust steam after cleaning is attractive but can be used with precautions and cannot be generalized.<br>• Excess steam can be used appropriately in the process to improve the energy efficiency of the drying process.<br>• Technoeconomic study of each configuration of the dryer combination is must. |
| 2. | Dibella et al. [178] | Integration of SSD with exhaust recompression | To recover the latent heat in the exhaust by using recompression of excess steam in exhaust for SSD Review of technical, economical, and commercial feasibility investigation of large range of products. | • The net energy reduction of 55% compared to conventional air drying is possible by SSD and thus equivalent proportion of reduction in emission of $NO_x$ and $CO_x$ due to reduced energy.<br>• By simple payback, economically feasible with 2 years of duration. |
| 3. | Shi-Ruo et al. [179] | LPSSD lab-scale test set-up | Silkworm cocoon Initial Moisture content 1.35–1.7 kg/kg dry cocoon/ experimental on lab scale /feasibility of SSD, drying kinetics and characterization of dried cocoon. 40–60 kPa gauge pressure. | • Drying time: Conventional methods @ 6 h.<br>• Optimal operation: Two stages drying- Primary drying of cocoon at 403 K with SS and then with hot air at 368 K secondary drying.<br>• Cocoon consumption reduced by 40–110 g cocoon/kg of raw silk in silk reeling. |

*(Continued)*

**TABLE 6.1 (*Continued*)**
**Details of the SSD with Applications and Features**

| S.N. | Paper Details with Authors | Study | Feedstock/Type of work/Features | Remark |
|---|---|---|---|---|
| 4. | Wimmerstedt [180] | Review of integration of SSD for fuel drying in thermal power plants or processes such as district heating, beet pulp, paper pulp, etc. | Integration of sugar beet pulp dryer using SS with sugar plant, drying applications: sludges, especially handling of landfill leachates, lumber, paper and paper pulp, and sugar beet pulp. | • Commercial SSDs of flash-and fluid-bed types are available on the market.<br>• Though SSD is not popular commercially, it will have good potential applications with the integration of excess steam. Even without steam reuse in certain applications, SSD is economical. |
| 5. | Fitzpatrick [181] | SSD process | Sludge drying/simulation on integration of SSD with sludge drying by anaerobic digestion | • Anaerobic digestion produces methane from sludge, can be dried using SSD.<br>• Thermal energy required for drying is generated by combustion of the methane produced by anaerobic digestion.<br>• The excess steam equal to the moisture evaporated can be utilized to maintain the temperature of aerobic digester, thus saving the energy by reuse, but more beneficial in cold climates ($<10°C$).<br>• Plant can be designed such that there will be no need of external supply of thermal energy but internally it will be balanced, i.e., thermal energy self-sufficient plant design is possible.<br>• No emission of dust and volatiles and potential for production of pathogens-free dry sludge product. |

(*Continued*)

**TABLE 6.1 (*Continued*)**
**Details of the SSD with Applications and Features**

| S.N. | Paper Details with Authors | Study | Feedstock/Type of work/Features | Remark |
|---|---|---|---|---|
| 6. | Aly [182] | Analytical simulation-SSD with recycle of steam integrated with two stage mechanical vapor compression (MVC) for compression of purged evaporated water with an intercooler and aftercooler. | Milk powder-concentration and drying processes | • SSD with recycle of supply steam appropriately and the excess vapor from the exhaust of drying process are compressed in a two stage compressor with an inter-desuperheater and after desuperheater.<br>• Gas turbine drives the system and its exhaust heat is recovered with the balance of energy being used to drive a conventional multiple-effect desalination (MED) to produce fresh water.<br>• Techno-economic analysis shows better performance than air drying in all respects with 3–4 years of payback time. |
| 7. | Namsanguan et al. [71] | Lab-scale SSD, Heat pump drying (HPD) and Hot air drying (HAD) | Shrimp/Comparison among two-stage drying configurations (i) SSD followed by HPD (ii) SSD followed by HAD and SSD/Optimal configuration and design conditions for best quality and lower energy. Drying time, temperature, quality attributes: shrinkage, color, texture (toughness and hardness), rehydration behavior and microstructure. | • SSD followed by HPD with tempering in between drying stages improve the drying rate in HPD and this hybrid drying combination gives better quality (degree of shrinkage, rehydration potential, color, textural properties, and microstructure).<br>• SSD followed by HAD, though noticed improvement in some quality parameters but no improvement in terms of shrinkages and rehydration. |
| 8. | Alves-Filho and Roos [183] | Reviews on hybrid drying, such as superheated steam dryers with evaporators | – | • Specific moisture extraction ratio (SMER) is 2.4–3.6 ($kg_w$/kWh), which is significantly higher than conventional dryers. |

(*Continued*)

**TABLE 6.1** (*Continued*)
**Details of the SSD with Applications and Features**

| S.N. | Paper Details with Authors | Study | Feedstock/Type of work/Features | Remark |
|---|---|---|---|---|
| 9. | Nimmol et al. [163] | Lab-scale low-pressure superheated steam and far-infrared radiation drying at the same time (simultaneous) | Banana/lab-scale test rig/effect of drying medium temperature and pressure on the drying kinetics, heat transfer, energy consumption; Comparison with combined far-infrared radiation-vacuum drying (VD-FIR) and LPSSD; | • Both configurations, LPSSD–FIR and VD-FIR are simultaneous drying processes.<br>• For each drying configuration, the drying rate reduces with increasing temperature and reducing drying chamber pressure, which is mainly due to increasing driving potential.<br>• On the basis of drying rates and the specific energy consumption at all tested processes, LPSSD–FIR at 90°C and 7 kPa was most optimal.<br>• Except 90°C and 7 kPa, drying time is lower than VD-FIR, indicating that inversion temperature will be in between 80°C and 90°C at 7 kPa. |
| 10. | Somjai et al. [118] | Lab-scale sequential hybrid drying SSD with HAD | Longan without stone- unpeeled with initial moisture content 300%–350% dry basis (25–30 mm 40 g without stone); Quality attributes-color, shrinkage, and microstructure. Drying temperatures for both SS (120°C–180°C interval of 20°C) and HA (60°C, 70°C) | • First SSD experiments, longan dried to the intermediate moisture content of about 200% dry basis (moisture content at the end of the first-stage SSD). This moisture content was selected from the preliminary observation on purely SSD of browning. In the second-stage hot air drying continued at temperatures of 60°C, 65°C and 70°C with an air velocity 0.7 m/s until the longan fruit reached final moisture content of 18% dry basis.<br>• Color for hybrid drying was improved, but shrinkage was increased with fewer pores compared to SSD.<br>• Hybrid, two-stage drying with superheated steam at 180°C followed by hot air at 70°C was the optimal. |

(*Continued*)

**TABLE 6.1 (*Continued*)**
**Details of the SSD with Applications and Features**

| S.N. | Paper Details with Authors | Study | Feedstock/Type of work/Features | Remark |
|---|---|---|---|---|
| 11. | Morey et al. [184,185] | SSD process | Distillers wet grains (DWG)/ simulation in aspen plus of integration of SSD with corn ethanol plant (biomass integrated gasification combined cycle (BIGCC))/ compatibility of SSD compared to HAD with steam tubes. | • Energy consumed 760–805 kJ/kg of water removed, and 1.3l of water recovered per liter of ethanol produced by SSD integration compared to energy consumption for the hot air operated steam tube dryer was 2660–2690 kJ/kg of water removed with no water recovery.<br>• SSD can save energy in co-product drying and reduce water use at biomass integrated gasification combined cycle (BIGCC) for generation of ethanol plant. |
| 12. | Aziz et al. [186–189] | Enhanced high energy efficient steam drying- Steam tube rotary dryer | Brown algae, low-rank coal, empty fruit bunch/simulation- commercial process simulator Pro/II (Invensys) | • Integrated steam drying system requires less energy compared to counterpart conventional hot air drying.<br>• Need to check practical feasibility of so many heat exchangers from engineering and process operational point of view. |
| 13. | Yamsaengsung and Tabtiang [190] | Pilot-scale hybrid dryer of SSD and HAD | Rubberwood boards/experimental on pilot plant/drying time and mechanical properties as static bending, compression strength, hardness, and shear strengths. Optimization parameters: temperature and velocity of SS and HA. | • Optimal hybrid drying is investigated for given rubberwood with conditions as 105°C–107°C of SS and 80°C HA. Drying time is reduced by 62% (to 64 h from 168 h)<br>• No effect compared to conventional for shear strength parallel-to-grain, but the mean compression strength parallel-to-grain was reduced by 24.2% and the mean Modulus of Rupture by 21.4%. However, the mean Modulus of Elasticity was increased by 30.4% and the mean of hardness by 16.4%.<br>• Overall energy consumption for the optimum hybrid drying process for given case was 11.3 MJ/kg of water. |

(*Continued*)

**TABLE 6.1 (*Continued*)**
**Details of the SSD with Applications and Features**

| S.N. | Paper Details with Authors | Study | Feedstock/Type of work/Features | Remark |
|---|---|---|---|---|
| 14. | Tolstorebrov et al. [174] | High-temperature heat pump for superheated steam drying integration using turbo-compressor | Pet food/analytical simulation integration of water refrigerant with turbo-compressor heat pump | • Analytical model using water as refrigerant shows good potential for proposed closed and open systems. |
| 15. | Atsonios et al. [175] | Fluidized bed dryer with SS (WTA dryer) integration with lignite plant/two configurations as (1). Excess steam from drying exhaust after recompression as heat source (2). Extracted steam from the turbine or extracted water from feedwater tank. | Lignite/drying technology integration in a modern steam cycle representing state of the art for a power plant firing low-rank coal. Optimization of existing lignite pre-drying concepts and their improvement in terms of overall plant efficiency. | • The fluidized bed dryer is a more efficient (electrical efficiency increased by 1.2%–2.1% compared to reference without any drying) option for retrofitting existing power plants.<br>• Steam recirculation and condensation in the dryer is the best option to provide adequate heat for drying.<br>• Higher plant efficiency for the optimized pre-drying process scheme and its integration with the overall steam.<br>• Feedwater heating requires more space in drying chamber, making fluidization more difficult. |
| 16. | Ratnasingam and Grohmann [191] | Commercial Kiln of 10 $m^3$, optimization of recipe (process parameters with schedule) for minimal drying defects and reduced drying time. | 50 mm thick rubberwood boards using a combination of superheated steam and hot air. Initial saturated steam at 100°C, followed by superheated steam at 110°C, and lastly hot air at 65°C. | • The rubberwood drying is improved considerably by mixed drying media with sequence as saturated steam then superheated steam and last hot air during the drying cycle.<br>• Optimal schedule in commercial kiln in Malaysia, drying time is reduced from 22 to 12 days for 50 mm thick rubberwood. |
| 17. | Johnson [192] | General SSD system | Analytical using Pinch and exergy analysis to improve energy efficiency for integrated systems | • SSD has extra 87% of recoverable energy from the condensing of excess steam, which can be integrated throughout an extended system.<br>• Steam dryer works across the pinch and easily integrate with into other systems. |

*(Continued)*

**TABLE 6.1 (*Continued*)**
**Details of the SSD with Applications and Features**

| S.N. | Paper Details with Authors | Study | Feedstock/Type of work/Features | Remark |
|---|---|---|---|---|
| 18. | Liu et al. [176] | Review of various evaporative dryers integration with lignite-fired power plant | Various dryers in literature for low-rank coal drying/survey cum review | • Compared to hot air dryers, SSDs are having higher efficiency (2%–6% electrical efficiency) due to more heat integration opportunity and exhaust steam heat conservation.<br>• Waste heat recovery and water recovery are added advantages for SSD. |
| 19. | Jaszczur et al. [193] | Lignite coal dryer (general term used) | Lignite coal with more than 50% Moisture (wet basis)/Integrated Gasification Combined Cycle for electricity production, coupled with a fuel preparation unit and a coal superheated steam drying system | • Integration of total seven sub-systems as fuel preparation (lignite drying), air separation unit (ASU), coal gasification, gas cleaning system, regeneration system heat recovery steam generator system, steam turbine and gas turbine for maximum performance.<br>• With SSD integration, efficiency will increase up to 48%–56% depending on the thermodynamic parameters, additionally reducing significant carbon emission.<br>• For this integrated process, investment is very low without much complexity. |
| 20. | Chantasiriwan and Charoenvai [194] | Integration of SSD with parabolic trough solar collector | Sugar cane juice drying/Simulation of cogeneration plant with SSD drying and PTSC, economic analysis | • Integration of cogeneration plant with SSD and PTSC increases the power output without changing fuel consumption rate.<br>• Simple payback period analysis proposes economically feasible investment. |
| 21. | Yao et al. [195] | Pilot set-up for Low-pressure superheated steam (LPSSD) and heat pump drying (HPD) | Fish (grass carp) initial moisture content (wet basis 70%–75%)/drying medium pressures on the temperature field, airflow field, drying time, equipment performance, the power consumption | • Heater is used to superheat the steam in addition to heat pump. Further application of heat pump is to condense the excess steam/ moist air maintaining the mass flow of the steam/moist air constant during the SS and HA drying.<br>• Due to the hybrid system, LPSSD with HPD during recirculation of steam at low pressure (7 kPa) energy consumption is lowest and increases with increasing pressure by high margin. It is highest for air drying. However, the drying also increases significantly. |

(*Continued*)

**TABLE 6.1 (*Continued*)**
**Details of the SSD with Applications and Features**

| S.N. | Paper Details with Authors | Study | Feedstock/Type of work/Features | Remark |
|---|---|---|---|---|
| 22. | Chantasiriwan [177] | Steam dryer for lignite-fired thermal power plant and thermal integration with air preheating | To investigate integration of steam dryer with heat recovery for air preheating in lignite-fired power plant/Simulation model<br>Extracted steam for feed water heating in feed water heaters and fuel drying in steam dryer, and uses heat recovered from steam dryer for air heating in heat recovery air preheater. | • The net efficiency of a thermal power plant can be raised by increasing the feed water temperature at the boiler inlet, increasing air temperature before combustion, or decreasing the moisture of fuel consumed by the boiler.<br>• Boiler efficiency increases by decreasing the minimum exhaust flue gas temperature without causing condensation of sulfuric acid vapor in flue gas.<br>• Simulation results show that the proposed power plant that supplies low pressure extracted steam to steam dryer to remove 20,000 kg/h of moisture delivers 0.88% more power output than the reference power plant.<br>• Overall, the integration of steam dryer and heat recovery air preheater, along with the installation of additional air heater, is technically and economically feasible. |

LPSSD using the lab-scale system. FIRD is used simultaneously with SSD/vacuum drying, i.e., at the same time, wet feedstock is exposed to superheated steam/vacuum drying and FIR. It was concluded that drying time for LPSSD-FIRD and vacuum-FIRD was lower than only LPSSD for all drying conditions using banana as feedstock. Between LPSSD-FIR and vacuum-FIR, the former required lower drying time for 90°C superheated steam temperature, indicating that, for combined drying rate, inversion temperature should be between 80°C and 90°C. From the specific energy consumption perspective, electrical energy required for two reasons, to maintain the vacuum and thermal energy supplied, reduced due to FIRD. Overall, from drying time and specific energy consumption, LPSSD-FIRD at 90°C superheated steam temperature and 7 kPa absolute pressure was the optimal configuration. However, the authors clearly specified that dried product quality was not investigated, which is the most important aspect to be studied further.

Another lab-scale work on combined dryers working in sequence (two-stage), SSD and HAD, for drying longan without stone was reported by Somjai et al. [118]. Both the dryers, superheated steam and hot air dryers (SSD+HAD), at lab-scale models are shown in Figure 6.1 [118] with important components. In this experimental work, both drying kinetics and product quality in terms of color, shrinkage, and microstructure were investigated. Also, two configurations of the dryers, SSD+HAD and only SSD, were compared based on these parameters.

It is observed that the overall drying rate of longan by combined SSD+HAD depends more upon hot air temperature than steam. Out of two configurations, the SSD and SSD+HAD gave better color. Higher temperatures for both steam and hot air medium gave better color, but shrinkage was lower in the case of SSD. Considering both drying time and product quality, two-stage drying with superheated steam at 180°C followed by hot air drying at 70°C was the optimal drying condition for drying longan without stone. Additionally, based on mass and energy balance, a simulation model was developed for the same and found to be close to

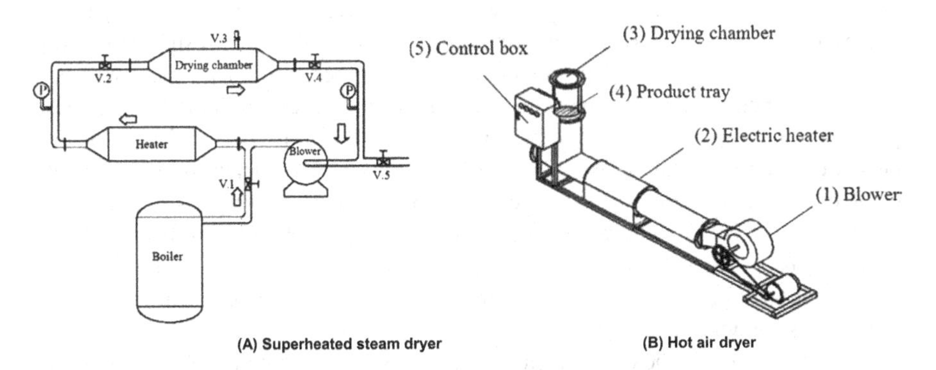


**FIGURE 6.1** SSD and HAD dryers used for the drying of longan [118]. (a) Superheated steam dryer. (b) Hot air dryer.

experimental results. This model is useful for predicting the moisture content, outlet temperature after drying, and drying rate under various conditions with relatively good accuracy for longan drying.

## 6.3 INTEGRATION OF SSD WITH PROCESSES FOR REUSE OF EXCESS STEAM

As discussed previously, the superheated steam is used to evaporate the moisture present in the feedstock, which becomes the part of the outlet steam at lower enthalpy. This excess steam, if used appropriately in the plant or its thermal energy is recovered, then overall efficiency of the drying plant increases significantly to the range of 55%–80%. So there are efforts for utilization of the excess steam by various means as follows:

1. Recompression of the excess steam for high-pressure applications or heat recovery. Sometimes, after recompression, this steam can supply compression and latent thermal energy to heat the recirculated steam, increasing the overall efficiency of the plant by 55% compared to conventional hot air dryers. The system's heat input will be in the range of 1500–2350 kJ/kg of water evaporation. This helps additionally to recover water as condensate (liquid water after condensation).
2. This excess steam at lower temperature and enthalpy (exhaust from the drying chamber) can be useful for the preheating of the feedstock or other process heating applications as per the availability.

### 6.3.1 Vapor Compression

The summarized important work from the literature on the recompression of exhaust steam in superheated steam dryer and suitable compressors till 1990 is given by Mujumdar [9]. Tolstorebrov et al. [174] developed and investigated high-temperature radial turbo-compressor operating on water as a refrigerant. Operating temperature range of the compressor will be a maximum of 255°C–269°C with suction pressure near the atmosphere to 4.6 bar. For pressure below the atmosphere, there will be a challenge of penetration of ambient air. For the reported case of 10,000 kg/h of product processed into dried pet food, the performance coefficient is 3.8 and drying specific energy consumption is 0.19 kWh/kg of water evaporation for efficient systems operated on electricity. For fuel/gas-based systems without exhaust recovery, these parameters are 2 and 0.37 kWh/kg. Overall the proposed energy efficient system is technically and economically feasible, but needs to run industrial trials for other issues to be evaluated.

Mujumdar [9] proposed various configurations of superheated steam with compression as shown in Figure 6.2.

The integration of the excess steam for process heating application can be implemented by using pinch analysis. A few configurations of reuse of excess steam are given in Figure 6.3 [9] with recirculation of drying steam required.

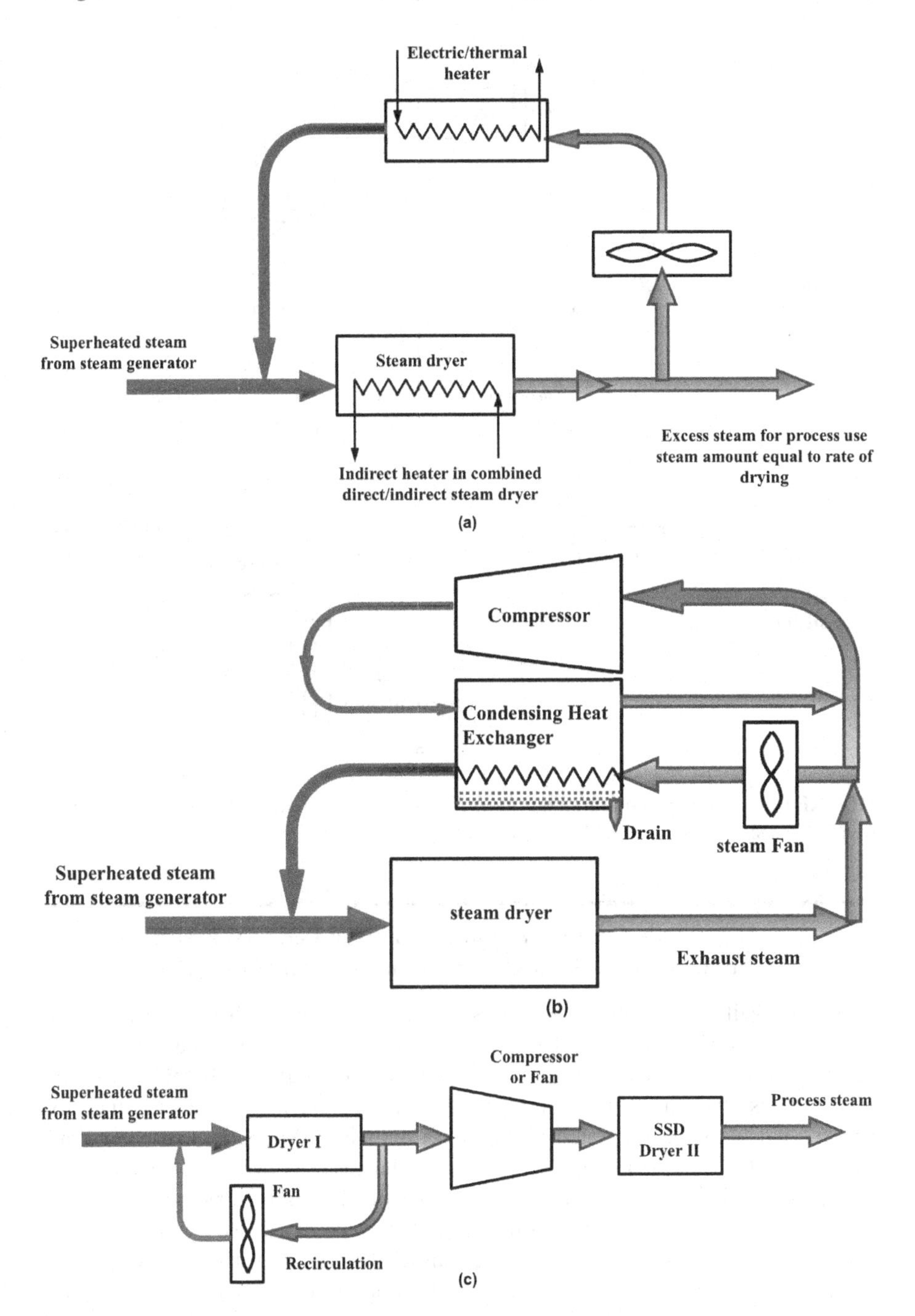


**FIGURE 6.2** Various configurations of reuse and recirculation of SS by compression and fan [9]. (a) SSD with mechanical compression of excess steam to add the thermal energy. (b) SSD with indirect heating by compression of excess steam supplying condensing heat to recirculating steam. (c) Two-stage drying with excess steam from the first stage after compressed used in the second stage.

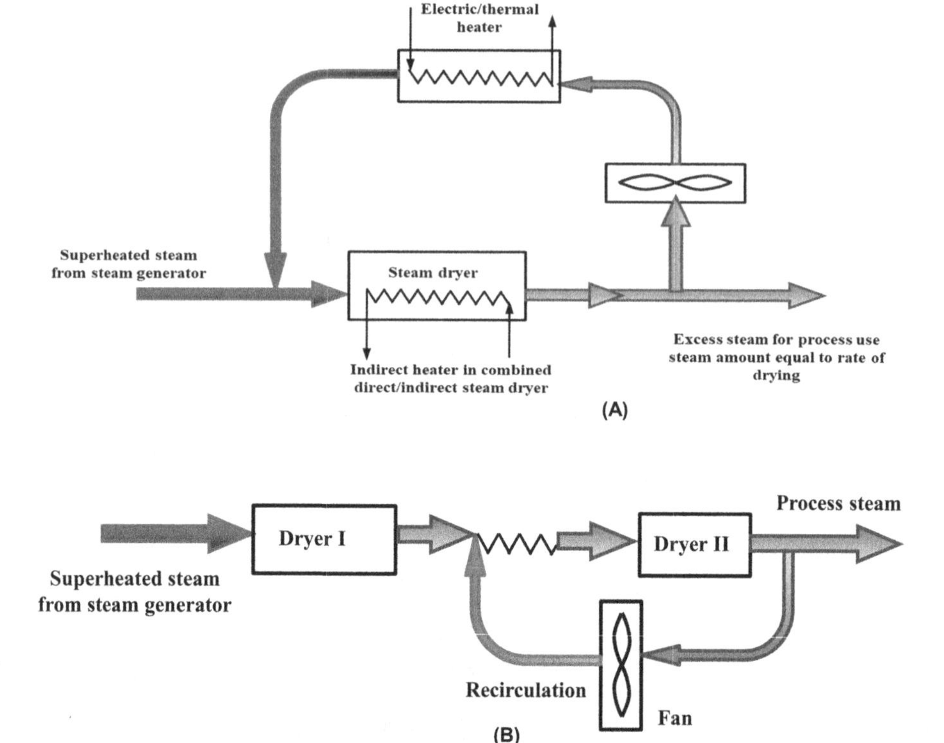


**FIGURE 6.3** Various configurations for the reuse of process steam by heating, recirculation, and process applications [9].

---

## Case Study 6.1: Integration of Fluidized Bed Dryer with Thermal Power Plant Operated on Low-Rank Coal [175]

Lignite is available abundantly and is used as a low-cost fossil fuel for power plants in several countries. However, due to high moisture content (more than 50% on wet basis), power plants have low efficiency, which can be improved by pre-drying lignite. Various dryers have been developed to dry lignite, but being an energy-intensive process, it needs to be integrated appropriately with the power plants to use low-grade heat available. This integration will definitely increase the overall efficiency of the plant. Liu et al. [176] reviewed the integration of evaporative drying technologies with lignite-fired power plants. In this case, low-rank coal (LRC) with a moisture content of 35%–60% (wet basis) is dried up to 10%–20% (wet basis) using superheated steam. They have summarized that the superheated steam dryers are better and enhance the electrical efficiency of the plant by 2%–6% by integration with lignite-based thermal power plant.

Atsonios et al. [175] had discussed the retrofitting pre-drying for existing lignite power plants. For superheated steam drying, two configurations were discussed based on the heating medium for the integrated dryer with power plant (sketch

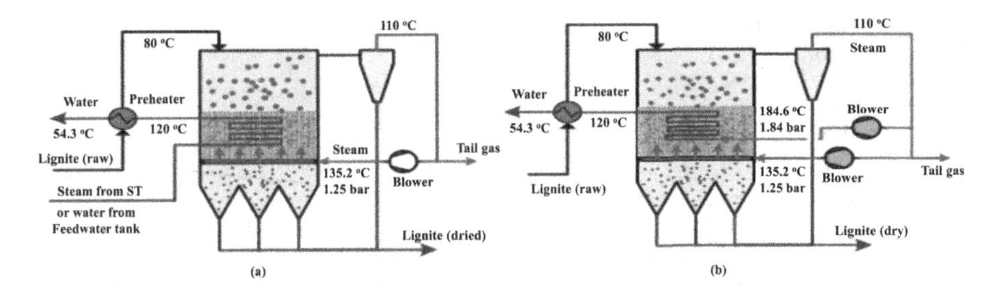


**FIGURE 6.4** Fluidized bed dryer (Wirbelschicht Trocknung Anlage (WTA) technology) with heat systems [175]. (a) Extracted steam from steam turbine or extracted water from feedwater tank. (b) Recompressed steam from the exhaust of drying chamber.

shown in Figure 6.4) as: (i) extracted steam from the steam turbine or extracted water from the feedwater tank; (ii) recompressed excess steam from exhaust of drying chamber.

In this case, for the configuration shown in Figure 6.4a, the extracted feedwater is taken at 190°C/12.62 bar pressure, and steam is taken from low-pressure stage of the steam turbine at 300°C/5.96 bar pressure (data not mentioned in Figure 6.4a). As shown in Figure 6.4a, the steam indirectly heats the medium (steam) in the fluidized bed drying chamber and condensate returns in the cycle through the water preheater before the feedwater tank. The evaporated moisture from the coal (fuel) is used as a heating medium, which saves the steam extraction. In another configuration shown in Figure 6.4b, the excess steam evaporated from the coal is first recompressed such that it can be reused both as a heating source and a drying medium. This system works independently (decoupled) from the rest of the plant and has no interaction and hence impact on the steam cycle. There might be a case that the energy required for recompression may be significantly high to match the heating energy. Overall, the fluidized bed dryers are efficient options and recirculation of steam with heat recovery from the excess steam will be an energy-efficient configuration showing 1.5% improvement in the overall plant.

It may be possible to have integration of steam dryer such that the exhaust temperature of flue gases will be reduced to minimum by heat recovery [177]. In this integration, the SSD is heated by bleeding steam from the low pressure turbine and the excess steam evolved due to evaporation of moisture from coal may be used for heating of the exhaust flue gas subsequently, reducing its temperature to minimum≈permissible value. The dried low-rank coal will be supplied to power plant. The flow diagram for the lignite-based thermal power plant is shown in Figure 6.5 [177].

For this drying process, rotary or fluidized bed dryers operated on superheated steam may be used. Overall, the integration of the superheated steam dryer and heat recovery air preheater with complete available heat conservation will improve power out from the plant and increase efficiency by 1%, making it economically feasible.

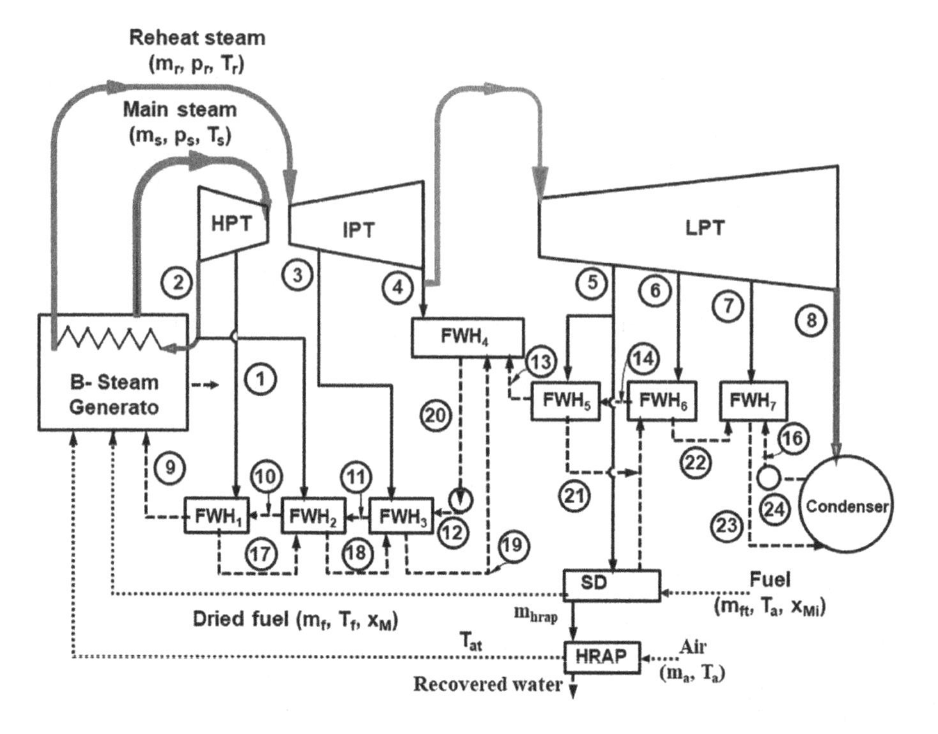


FIGURE 6.5 Lignite coal-based thermal power plant [177].

## 6.4 CLOSURE

In general, the integration of SSD in the process is an evolving area of study, where optimizations of the system's parameters need to be addressed appropriately for each configuration of dryer. It may be possible to use tools such as pinch analysis to optimally design the system better. As the overall system operation will become complex, it is important to check operability and controllability in addition to the product quality and energy savings.

Further, in combined drying systems, recently few works at laboratory scale were performed indicating its potential. In the coming years, more studies are expected to be performed to combine the best features of the different dryers so that the drying process will be optimal. The best thing about the superheated steam drying is its high drying rate, which results in low drying time if the temperature is above inversion temperature for that case. Further, it improves product quality; however, due to the higher temperature of the product, there are limitations due to browning or burning of drying material, which can be overcome by combining it with other suitable drying technology.

# 7 Modeling of SSD and LPSSD

## 7.1 IMPORTANCE OF MATHEMATICAL MODELING IN SSD

As discussed previously in this book, the superheated steam drying (SSD) technique is an energy-efficient technique that can provide very high-quality products. We also discussed other advantages such as a smaller carbon footprint and recovery of substance lost during the drying process, although there are certain limitations such as complexity in design, start-up/shut-down process, cost involved, and the inability of its use when a product requires oxidation [9,26,28]. The SSD technique has been successfully used in various dryer types, including both batch and continuous. There have been several experimental research studies carried out at different scales to understand the working principles and other aspects of SSD technique [26,28]. However, only experimental studies are insufficient to explore and understand multi-faceted drying technique like SSD, hence several studies on theoretical modeling and numerical simulation of SSD technique have been carried out. In general, the mathematical modeling studies in drying focus on modeling of drying kinetics or modeling of the drying system by solving complex heat and mass balance equations. The theoretical research studies of SSD have focused on drying kinetics, understanding physical phenomena in single particle drying, design of selected dryers (e.g., packed bed and fluid bed dryers) using appropriate heat and mass balance equations, product quality, and computational fluid dynamics (CFD) approach for various applications. This chapter is a quick overview of several theoretical studies of SSD.

## 7.2 SINGLE PARTICLE DRYING USING SSD

The important objective of theoretical study of single-particle SSD is to understand the influence of the particle structure on drying behavior using superheated steam (SS). As discussed in Section 2.5 of Chapter 2, SSD process consists of three stages, condensation of water on the product surface, surface evaporation (pseudo constant drying rate), and the falling rate. There has been extensive experimental and theoretical work carried out to understand the complex interaction between the drying medium used and the solid particles dried and study heat and mass transfer phenomenon [196–202].

The researchers have used various approaches based on different hypotheses. These include modeling based on the first principles, continuous/continuum models, simplified lumped models, and empirical/semi-empirical reaction engineering approach (REA). Each of these approaches requires several assumptions such as

DOI: 10.1201/9781003275299-7

uniform properties, constant thermo-physical properties, assumption on the way diffusion occurs. Traditionally used simple mathematical approaches to model drying kinetics such as Page model, effective diffusivity-based models are not discussed in this chapter.

An extensive review of mathematical modeling of single particle drying using SS reveals a significant improvement in the mathematical approach to predict general drying behavior, the moisture content and drying rate, predict particle temperature and the shrinkage phenomenon. Although the majority of research studies use momentum, heat, mass transfer equations in the mathematical approach, the difference lies mainly in the way the movement of moisture within the particle is accommodated, how the initial moisture condensation is considered, the method of including shrinkage into account. Table 7.1 summarizes the approach and outcome of selected theoretical research studies on single-particle drying using SS.

Most studies start with a description of a physical model used for their theoretical study, which is a replica of experimental setup if the comparison is made with experimental data. This is followed by writing necessary constitutive equations based on their approach. Then one requires properties of SS and the solid particles used for their theoretical study. The basic information needed about steam is temperature and steam pressure. However, extensive information about solid particles is needed to better understand and predict the drying phenomenon, such as particle structure, moisture content, initial temperature, and other physicochemical properties. The numerical approach used depends on the complexity of the mathematical models used [196–204]. Using appropriate initial and boundary conditions, the set of mathematical models is solved to mostly predict changes in moisture content and/or particle temperature. This information is compared with experimental data to validate the goodness of the approach employed. Most researchers used appropriate experimental rigs to collect and compare experimental data. However, there are a few studies that focus only on theoretical approach to better understand the use of SS. In general, the mathematical models developed for single-particle systems are used to study the effect of operating parameters on various outputs of interest and serve as a model for mass transfer for simulation of SSD of multi-particle systems [196–204].

## 7.3 MODELING OF PACKED AND FLUIDIZED BED DRYING WITH SS

As discussed in the introduction chapter, SS is a drying medium used to replace the use of conventional hot air, and it can be used in any convection drying system that has traditionally used air as the drying medium [9,26,28]. Hence, SS has been used in various types of dryers, including tray dryers, rotary dryers, variants of fluidized bed dryers, pneumatic conveying dryers, and packed bed dryers. In Chapters 3–6, we have discussed several applications of integrating SS with different conventional and advanced drying techniques for applications from biological products to minerals. In most of these applications, the product characteristics necessitate the use of either a packed bed or a variant of fluidized bed. Hence, there has been some theoretical work carried out to understand the drying behavior in these dryer types [205–212].

**TABLE 7.1**
**Summary of Selected Theoretical Work on Single Particle Drying Using Superheated Steam**

| S.N. | Details of the Research Article | Study | Approach | Key Findings/Remarks |
|---|---|---|---|---|
| 1. | Hager et al. [196] | Mathematical modeling of drying of single porous sphere of ceramic. | Foundational transport equations are solved with only one adjustable parameter. The required transport coefficients were either measured or calculated theoretically from the pore size distribution. Darcy's law is applied to both liquid and gaseous phase. Mass and energy balance equations were then solved using appropriate initial and boundary conditions. | • Simulation results suggest overprediction of drying rates at the start of superheated steam drying.<br>• Authors concluded that it could be a result of high estimated capillary movement because of an assumption that each pore same size.<br>• Authors also mentioned another possible reason for overprediction which could be inability to take local internal equilibrium into consideration. However, the internal temperature curve is well predicted by the mathematical model. |
| 2. | Chen et al. [197] | Development of single particle model to simulate the drying of granular media in superheated steam. | A mathematical model for drying liquid droplets in superheated steam was first successfully developed to validate the phenomenon that occurs during the initial condensation leading to an increase in size of the droplet. Authors then developed a mathematical model for drying of porous particle to include initial condensation phenomena that occurs in SSD. Authors considered the idea of receding front and Darcy flow of vapor. The other motivation was to investigate inversion-temperature. | • For drying of droplets, the measured mass was plotted as a function of time, while for single particle drying, particle temperature and evaporation rate was plotted.<br>• The experimental data was compared with the model prediction, which matched very well. Authors even took radiation effect into account for steam drying, although the effect of radiation was found to be negligible.<br>• After validation, the model was used to study the parametric effect. The time required to reach steady state was found to depend on permeability and size of particles.<br>• The initial condensation is also a strong function of initial particle temperature. For droplets, the inversion temperature was found to decrease with increase in the gas flow rate and initial droplet size, while it goes through a minimum for porous particle. |

(*Continued*)

**TABLE 7.1 (*Continued*)**
**Summary of Selected Theoretical Work on Single Particle Drying Using Superheated Steam**

| S.N. | Details of the Research Article | Study | Approach | Key Findings/Remarks |
|---|---|---|---|---|
| 3. | Hosseinalipour and Mujumdar [198] | Superheated steam drying of a single particle in impinging stream drying | Computational fluid dynamics simulation of 2-dimensional turbulent opposing jets on superheated steam drying of single particle was studied. Several turbulence models were tested and the effect of variables on particle movement and drying behavior of single particle was studied. | • The flow, heat and mass transfer simulation study showed promising results, although comparison of simulated results with experimental data was not presented. Different k-ε models were used to predict turbulent flow and heat transfer for various flow characteristics in the opposing jet impinging stream drying.<br>• For the specified particle characteristics, the residence time was found to vary significantly depending on the position of the release of particle, this in turn changes the amount of moisture removed in a given length of the dryer. This is an interesting study showing promising results on SSD of particle in impinging stream dryer. |
| 4. | Kiriyama et al. [199] | Experimental and theoretical study on SSD of lignite particle | Authors used a simulation model by assuming coal particle to be isotropic. A particle of assumed size of 30 mm was divided into 61 layers (concentric spherical shells). Transfer of heat, evaporation, transfer of water and shrinkage was considered in these segments. Heat transfer to the surface and inside the particle were modeled appropriately. Transfer of free water was assumed to be by capillary pressure gradient, diffusion by vapor pressure gradient. To simulate shrinkage of lignite particle (volume change), the volume of steam is assumed to decrease during drying. A linear shrinkage is assumed | • The approach used by authors has allowed good prediction of the moisture and temperature of particle.<br>• The experiments and simulated results showed good agreement for moisture content, particle temperature and drying rate for the temperature range of 110°C–170°C. |

*(Continued)*

**TABLE 7.1** (*Continued*)
**Summary of Selected Theoretical Work on Single Particle Drying Using Superheated Steam**

| S.N. | Details of the Research Article | Study | Approach | Key Findings/Remarks |
|---|---|---|---|---|
| 5. | Le et al. [202] | Superheated steam drying of cellular plant media using continuum scale modeling with potato as a reference material | A continuous model to describe heat and mass transfer inside cellular plant porous media was developed for drying using SSD at atmospheric pressure. The module considers advective liquid and vapor flow inside the intracellular void space, while diffusion of liquid across the cell membranes. Linear shrinkage model was assumed to model the surface shrinkage. | • Model assessment shows that the experimental data matches well with the simulated results. The deviation is a bit more for certain drying temperatures which authors think could be a result of using same value of certain properties for all simulations.<br>• Although the model predictions are good, authors suggested some improvements in the approach as also mentioned the complexity of the model which may not allow its use for multi-particle system. |
| 6. | Le et al. [203,227] | Development of characteristics drying curve model from continuum model simulations for SSD of single wood particles | The continuum scale model describing heat and mass transfer for superheated steam drying of single wood particle at atmospheric pressure was developed using the properties of particle obtained experimentally. The continuum model was then reduced to a lumped model (characteristic drying curve model, CDC). The model parameters predicted using simulation results from continuum model were used as input for the lumped model. | • The continuum model was validated with experimental data and used to understand parametric effects and to establish correlations for the parameters appearing in the simplified lumped model.<br>• Unlike the conventional experimental method of finding the parameters needed for characteristic drying method, authors used continuum model to find those parameters.<br>• Authors found out that intense drying conditions have significant impact on parameters like critical moisture content. The numerical result obtained from characteristic drying curve model compared well with experimental data. Authors propose use of CDC model in CFD simulations. |

(*Continued*)

**TABLE 7.1** (*Continued*)
**Summary of Selected Theoretical Work on Single Particle Drying Using Superheated Steam**

| S.N. | Details of the Research Article | Study | Approach | Key Findings/Remarks |
|---|---|---|---|---|
| 7. | Hao et al. [200] | Modeling of moisture and heat transfer during SSD of wood drying considering migration of evaporation interface | The phenomena and the associated parameters of the evaporation interface were integrated into the appropriate governing equation(s). This helps authors to characterize the migration of the evaporation interface during the SSD of wood. | • The heat and moisture transfer model with evaporation interface migration was able to accurately explain the superheated steam drying and predict the dynamic changes in temperature, moisture content, interface evaporation rate, volume evaporation rate, water vapor density, and relative humidity at any location of the wood specimens during the drying process.<br>• The study also helped to gauge the possibility of using Fick's diffusion equation with variable diffusivity to characterize the Darcy flow. This reduced the number of equations and physical parameters needed. |

In a typical theoretical study on fluidized bed SS, the focus is on understanding the particle behavior/movement, characteristics of fluidization, residence time of particles, and the other parameters like temperature and moisture content. The information needed for simulations is typically the particle size distribution of the feed, particle characteristics, the moisture content, and SS properties. One of the earliest works was carried out by Heinrich et al. [205], which focused on drying of granular solids using SS in a batch fluidized bed. The mathematical approach consisted of writing heat and mass balance equation for different mechanisms in the three phases namely condensation, constant rate, and falling rate period. The authors then transformed a system of differential equations into dimensionless form and described with dimensionless characteristic numbers derived from the similarity theory. The simulated results were validated with the experimental findings from SS fluidized bed drying of test material (Seramis granules) for different solid batch sizes, and various steam inlet temperatures and mass flow rates. Besides this, the authors showed industrial applications of their theoretical study by applying the findings to roasting of coffee beans.

Taechapairoj et al. [206] carried out a similar study to understand the temperature and moisture behavior of paddy in a SS fluidized bed dryer. The mass and energy balance equations were written separately for initial condensation phase, constant rate drying period, and falling rate phase. The kinetics of gelatinization of paddy under fluidization condition was described by zero-order reaction and was included in the simulations. A single algorithm was used to check if the particle temperature is beyond the steam temperature to decide if the initial condensation phase is complete. While moisture content criterion was used to switch for theoretical analysis from constant to falling rate period. The predicted moisture content and particle temperatures were compared fairly with the experimental results for a wide range of operating conditions. The authors then used the model to study the effect of variables like bed depth, superficial velocity, and temperature of SS. Kovenskii et al. [207] carried out a simplified theoretical study for drying of word particles in fluidized bed. The aim of this study was to determine the influence of different physical, hydrodynamic, and thermophysical factors on drying. The authors assumed only the falling rate phase of drying and that the temperature of escaping steam is equal to particle temperature. The fluidized bed was considered to have an idea mixing with particle heating being gradient-free. The heat and mass balance equations were written based on the above-mentioned assumptions. The authors found satisfactory agreement between the calculated and experimental data on the mean moisture content of particles as a function of the drying time.

Chen et al. [208] integrated their single-particle drying model with a two-phase hydrodynamic model to simulate the continuous drying of coal particles in a fluidized bed using SS. They made general assumptions such as the existence of bubble and dense phase, well-mixed solids in the dense phase, well-mixed gas in the dense phase (hence properties of gas are spatially uniform), and plug flow of gas in the bubble phase. Energy and mass balance equations for initial condensation and drying period, along with the sorption isotherm model, are used to incorporate variation of particle temperature and moisture content with residence time, the effect of initial condensation, and the effect of variation in the steam flow. The effect of particle size and operating conditions on the performance of a fluidized bed dryer was evaluated. The model was also extended to study the effect on inversion temperature.

Unlike the fluidized bed dryer, where the gas and particles are assumed to be constant throughout individual phases, the properties vary along the length of the packed bed. Hence, the theoretical study requires solving a set of partial differential equations. Tang et al. [209] carried out a theoretical analysis of SSD of a fixed bed of Brewer's spent grain. The energy and mass balance equations were formulated like hot air drying. While for the kinetics, the drying rate equation was obtained from a thin-layer drying model determined from a previous study by the authors. The mass flow rate of steam was used as one of the variables for steam drying. By making reasonable assumptions as minimum change in enthalpy of drying medium in small elemental bed volume, and by eliminating heat transfer coefficient and specific area, the fixed bed drying model was simplified. To analyze the drying process, the fixed bed was divided into three zones, dry, drying, and wet zones. The depth of the dry zone increases while the depth of the drying and wet zone decreases as the drying progresses. The simulation results were in reasonable agreement with the experimental data, barring the final drying phase. Tran [210] carried out modeling of drying in packed bed by SS using a continuum scale model. A reaction engineering approach (REA) model was built to describe mass transfer between particle and steam. The dehydration of porous media is described as a reaction which needs the activation energy to overcome energy barrier. After validating the single particle drying model, it was combined with energy and mass balance equations written for the packed bed. The mathematical equations were solved to understand the changes in moisture content, particle temperature and steam temperature during the drying process.

Most of the other studies on mathematical modeling of SS packed bed drying use similar approaches with minor variations in the way drying kinetics are modeled or how the constitutive equations are modified or simplified. Messai et al. [211] used a one-dimensional mathematical model to describe heat and mass transfer during the drying of a packed bed of porous particles with SS and air. Their model is based on the scale-changing approach. For SSD, new correlations were developed to estimate parameters used to quantify mass flux, while for humid air drying the expression for mass flux was deduced from literature. Sghaier et al. [212] also carried out similar work but their model is based on the averaging approach using two changes of scale. They assumed a convective heat exchange between the particle bed and steam. A two-temperature macroscopic model was used to describe heat transfer between the porous particle and fluid phase. Drying kinetics deduced from a single-particle model from their previous study was used to introduce mass transfer.

## 7.4 CFD APPROACH IN SSD

CFD approach has been used successfully to understand the drying process and enhance process design and development [28,213]. CFD studies offer a detailed understanding of very complex fluid flow, heat and mass transfer phenomena occurring within the drying process, hence it is an excellent tool for visualization of various parameters such as moisture, temperature profile inside solid material to be dried, the particle movements/trajectories, distribution of drying medium and its properties inside the dryer, and other useful information. Once a CFD model is validated successfully with experimental data, it is an excellent and cost-effective tool in product development and scale-up. There have been several studies showing the usefulness of CFD to carry out parametric studies without a need to do experiments, such as spray drying, fluidized bed drying, and impingement drying.

Usual steps involved in the CFD analysis include deciding the physical model → choosing the framework for the multi-phase system → defining/collecting information for the initial and boundary conditions → formulating the governing equations relevant to the system and additional constitutive equations needed → collecting properties of drying medium and wet solid particles → selecting if 2-dimensional or 3-dimensional simulations are planned → geometry and mesh generation → using appropriate numerical schemes for discretizing the governing equations and solving the resulting algebraic equations → validation of results → post-processing of results and analysis [213].

An extensive literature review shows that there are numerous research studies on CFD analysis of SSD of particulate matter using various drying techniques. Most of these studies focused on single-particle drying. However, there are a few studies focusing on SSD using specific techniques. This includes SSD in fluidized bed dryer, and spray dryer [45,214–219]. Table 7.2 summarizes selected CFD studies on SSD.

## 7.5 MATHEMATICAL MODELING OF LOW-PRESSURE SUPERHEATED STEAM DRYING

The need for the use of low-pressure superheated steam drying (LPSSD) and its operating principles have been discussed in detail in Chapter 5 of this book. The use of low-pressure steam allows exploration of various benefits that SS offers, and at the same time, it ensures the product quality is not affected for the heat-sensitive materials [9,26,28]. Although there have been several experimental research works reported in the literature on the application of LPSSD, limited theoretical studies have been carried out. One of the earlier studies was carried out by Suvarnakuta et al. [220] in 2007. A simple mathematical model was developed to predict the temperature and moisture profile of carrot cubes undergoing drying using a low-pressure superheated steam. The carrot cube was assumed to be isotropic and homogenous, and the initial condensation stage was neglected. The mass transfer within the carrot was assumed to be controlled only by diffusion. The shrinkage was considerable and was considered as uniform shrinkage in the model. Heat transfer within the carrot cube is driven by conduction as temperature gradients developed in all directions. The selected modeling approach was simulated using COMSOL Multiphysics software. The moisture ratio and temperature were used for the comparison of simulated results and experimental outcomes, and a good match was found between the two. The authors suggested the need for several improvements in their approach and later proposed a revised modeling approach for LPSSD of the same product. In their later work, Kittiworrawatt and Devahastin [221] proposed a revised model that considers the effect of initial steam condensation in terms of film condensation. They also used more realistic mass transfer boundary conditions in terms of the vapor pressure gradient and the physical condition at the drying surface. They continued to include

**TABLE 7.2**
**Summary of Selected Theoretical Work on Computational Fluid Dynamics Study of Superheated Steam Drying**

| S.N. | Details of the Research Article | Study | Approach | Key Findings/Remarks |
|---|---|---|---|---|
| 1. | Ramachandran et al. [214] | Three dimensional CFD modeling of SS drying of single distillers' spent grain pellet | A 3-D model of the pellet and drying chamber was created. The governing equations (Reynolds-Averaged Navier-Stokes equations) were solved using the finite volume method and SIMPLEC algorithm within the CFD package (ANSYS CFX). | • The simulations result showed an excellent match with experimental results with less than 10% error.<br>• The moisture content and the surface temperature of the pellet as a function of drying was studied.<br>• Initial condensation phenomenon was also predicted well.<br>• The moisture distribution inside the pellet as a function of time was also visualized theoretically.<br>• Authors concluded it as a useful tool for design and optimization of large-scale SS dryer. |
| 2. | Ramachandran et al. [45] | CFD study of superheated steam drying of compacted distillers' spent grain coated with solubles. | Similar approach to Ramachandran et al. [214] was used. However, this study involved multi-layered/coated product with wet material dried over relatively dry core. Hence, different diffusion models were used. | • The match between model prediction and experimental data was better.<br>• Sensitivity analysis showed more prominent effect of SS temperature on drying time than the SS velocity.<br>• Moisture profile of coated pellet was studied and visualized |
| 3. | Kimwa, et al. [215] | Computational modeling of moisture content and temperature distribution in corn in superheated steam drying | The simulation used a cone geometry to represent the corn kernel. The Eulerian-Eulerian approach was used to model multiphase flow as the fluid and solid phases are interpenetrating/interacting. Corn is assumed to be homogeneous, isotropic and no shrinkage effects were considered. Internal heat and mass transfer is defined by diffusion. | • The CFD simulated/predicted experimental data for corn kernel was compared with a similar data for other product (experimental data from other study on dying of distillers' spent grain [45])<br>• Authors found a decrease in the duration of initial condensation as a higher temperature and/or velocity of superheated steam is used.<br>• The variation in moisture content and surface temperature as a function of time was studied.<br>• Authors suggested a more focused investigation that considers a study on the change in the corn's physical dimensions and geometric position with respect to the superheated steam. |

*(Continued)*

**TABLE 7.2 (*Continued*)**
**Summary of Selected Theoretical Work on Computational Fluid Dynamics Study of Superheated Steam Drying**

| S.N. | Details of the Research Article | Study | Approach | Key Findings/Remarks |
|---|---|---|---|---|
| 4. | Frydman et al. [216] | Numerical simulation of a spray dryer using the computational fluid dynamics (CFD) carried out to compare use of hot air and superheated steam. | A discrete droplet model was used to solve the partial differential equations of momentum, heat, and mass conservation for both gas and dispersed phase. Simulations were carried out for a pilot scale spray dryer operated using air and superheated steam. | • Authors find almost no difference in the flow pattern when using superheated steam or air as a drying medium, except for the recirculation when using steam.<br>• The general behavior of droplets in air and steam is same. The droplets evaporate faster.<br>• The problem of deposition of particles was found to be more pronounced in air than steam.<br>• Under the conditions used for their theoretical study, a higher volumetric drying rate was observed in superheated steam drying. |
| 5. | Frydman et al. [219] | CFD simulation of spray drying using superheated steam. | The model describes momentum, heat and mass transfer between discrete droplet phase and continuous steam (gas phase) using a finite volume method. Shrinkage during drying is also accounted for in the model. | • The CFD model is validated by comparison with experimental data on temperature inside the spray drying chamber.<br>• Authors studies the gas flow patterns and particle trajectories inside the spray drying chamber used for the simulations.<br>• The temperature field inside the chamber was calculated.<br>• Simulations confirmed the feasibility of using SS for spray drying and allowed understanding the drying phenomena. |

(*Continued*)

**TABLE 7.2** (*Continued*)
**Summary of Selected Theoretical Work on Computational Fluid Dynamics Study of Superheated Steam Drying**

| S.N. | Details of the Research Article | Study | Approach | Key Findings/Remarks |
|---|---|---|---|---|
| 6. | Xiao et al. [217] | CFD modeling and simulation of superheated steam Fluidized bed drying process with experimental data from drying of rapeseeds. | Authors used the Eulerian-Eulerian multiphase model to describe vapor-solid two-phase turbulent flow. For heat and mass transfer model, the drying process is divided into three regions, condensing and heating period, constant drying rate period, and falling drying rate period | • The main objective of this study was to compare the simulated results with the experimental findings from the fluidized bed rig.<br>• The simulated drying data was compared with experimental drying data for wet rapeseeds, which was mostly in good agreement except the initial condensation period where relatively more deviation is found. |
| 7. | Mohseni et al. [218] | Analysis of biomass drying in a vibrating fluidized bed dryer with a DEM CFD simulation tool | Approach involves resolving particles as discrete elements coupled via heat, mass and momentum transfer to the surrounding gas phase leading to Lagrangian-Eulerian coupling approach. The authors have used OpenFOAM as the computational tool. Experiments were carried out in industrial-scale vibrated bed dryer to estimate the moisture content, density, and size distribution of materials and residence time for comparison with simulation results. | • The moisture content at different instances and total residence time showed good agreement between experiment and simulation results.<br>• Theoretical results showed that higher temperature, velocity of inlet gas, as well as initial dryer temperature increase the drying rate and decrease the final moisture content of the materials.<br>• As expected, higher inlet moisture content of biomass requires longer residence time.<br>• Simulation results also include visualization of temperature distribution for gas and particles. |

the effect of shrinkage which was significant. The revised model predicts the center temperature and average moisture of the carrot sample mostly well, except at very high temperatures and lower pressures.

Elustondo et al. [222] carried out mathematical modeling of moisture removal rate from foodstuffs (shrimp, banana, cassava, apples, potatoes slab) exposed to LPSSD. They used a reduced model to calculate drying rate, which assumed that water removal is carried out by evaporation in a moving boundary, which means the water vapor needs to travel through a dried layer built because of drying. The theoretical model developed features dimensionless parameters to allow for the influence of form, shrinkage effect, and boiling point rise. The dimensionless drying rate predicted using the model of all the foodstuffs materials matched well with the experimental data. Shrivastav and Kumbhar [113] studied the selection of the analytical models and development of artificial neural network (ANN) models for LPSSD of paneer. In addition to the Page model (which typically fits the drying kinetics for almost every material), they used the ANN approach. The input layer in their model had two nodes corresponding to two processing conditions (drying time and weight change corresponding to time). The output layer consisted of three neurons or dependent variables (moisture content, drying rate, and moisture ratio). The prediction of experimental data using their theoretical approach was very good.

## 7.6 CLOSURE

In this chapter, a overview of theoretical approaches used for selected scenarios in the use of SS for drying is provided. This included single particle drying, fluidized and packed bed drying, CFD approach, and LPSSD. Besides this, there are several other aspects of SSD that researchers have explored using mathematical modeling and simulation. This includes a study conducted by Pang [223] where the mathematical model for high-temperature air drying was modified and extended for SSD. The key changes for SSD were external heat and mass transfer process and the calculation of equilibrium moisture content. The external mass transfer coefficient in the SSD was found to be much higher than hot air drying, while heat transfer coefficient was comparable. Adamski et al. [224] carried out mathematical modeling and simulation of SS flash drying of tobacco. A two-dimensional model of pneumatic transport consisting of momentum, heat, and mass transfer was used with solid represented by single isometric particles of the same size, while no radial distribution of velocity, temperature, or concentration in the pipe is expected. Drying rate was determined using a sorption-equilibrium equation obtained by fitting the experimental results. The predicted data was successfully compared with data from a large-scale flash dryer. Hamawand et al. [225] studied the drying process of banana slices with SS. Like many researchers did, the modeling of the drying process was divided into three periods, viz., initial condensation, constant rate period, and falling rate period. The net rate of evaporation or condensation per unit droplet surface area was governed by modified Hertz–Knudsen equation. Separate equations for drying rate in constant and falling rate periods were written.

The temperature and moisture content predicted using the model were in close agreement with the experimental data. Mathematical models can also provide insights into how different process parameters affect the quality of the dried product.

There are certain studies focusing on the quality of dried products. These include work by Fu et al. [226] to study the effect of deformation on heat and mass transfer in the drying of potato chips, and Junka and co-authors' [108] work on the drying of decelerated rancidity of treated jasmine brown rice. Besides these, there are some studies focusing on the modeling of SSD using the empirical or semi-empirical approach. In general, a review of various theoretical approaches used for SSD was carried out. Although we have not included the mathematical models in this chapter, we have tried to provide a synopsis of major assumptions, approaches, and key findings, which we feel can serve as useful guidelines for the authors. The theoretical approach certainly helps to understand the SSD techniques in greater detail and helps scale-up and optimization of the drying system, hence we certainly will see more advanced studies in this field to address other dimensions of SSD.

# 8 Advances, Prospects, and Global Market Demands *SSD*

## 8.1 INTRODUCTION

Superheated steam drying (SSD) technology has lots of potential but is commercially available only in few areas such as wood, beet pulp, biomass, and certain food, vegetables, and fruits items. Even though they have huge and vital merits, vast applications are still left unexplored owing to the appearance of certain issues (listed in Chapter 1 and a few in this chapter) during utilization. Thus there needs to be continuous advancement of technology to the next stage, addressing the challenges appropriately. Furthermore, it is worth discussing sustainability aspects due to the huge energy consumption and the importance of the green footprint of the product and process. In this chapter, sustainable perspective and future prospects with global market demands of SSD are discussed.

## 8.2 SUSTAINABILITY ASPECTS OF SUPERHEATED STEAM DRYING

Sustainability of processes is efficient energy consumption, generating no or minimum possible waste, and no or less emission of gases affecting global warming. Drying technology is not only to preserve but also to generate dried items with certain characteristics. So in the sustainability of drying technology, in addition to energy and environmental aspects [9], good-quality product with the highest level of required characteristics (for food items, nutrient retention and flavor together with microbial safety) are also important to be considered [228].

Generally, in a thermal drying process using hot air drying media, up to 40% of the total energy supplied is utilized for the evaporation of moisture present in the feedstock, which ends up in the exhaust gas emission at low grade in addition to energy lost. This low-grade energy is uneconomical for complete utilization though a lot of research is going on to recover as much as possible by recirculation, heat pump, and other means. In the case of recirculation of exhaust gas, part of it needs to be sent to the atmosphere due to the build-up of moist air for conventional hot air drying. On the other side, being superheated steam as a drying medium it is possible to have appropriate utilization of exhaust steam as a closed cycle and surplus steam utility for other applications in the same process or supplementary systems. It can be recirculated back into the system after cleaning and appropriately processing (as discussed in Chapter 6), reducing the energy requirement and surplus steam equivalent to the amount of moisture evaporated can be consumed for preheating or other applications

DOI: 10.1201/9781003275299-8

in the same process or supplementary systems [9]. If there is no application as stated above, then it can be condensed to get hot water. Effectively, there is no emission from the SSD, which further reduces odor and other pollutants evolved during drying [3]. Few related literatures are cited in Table 8.1. Overall, the optimal design of the drying plant with a superheated steam medium will have excellent sustainability.

Romdhana et al. [3] reviewed pilot and industrial superheated steam dryers from the perspective of energy efficiency. In this work, lab-scale and pilot SSD such as fluid bed, fixed bed, kiln, impingement jet, and flash drying given in literature and existing industrial SSD manufacturers like Danisco, NIRO, BMA, EnerDry (supplier of fluid bed); Bertin Promill (supplier of band dryer); Eirich GmbH & Co. KG (supplier of Rotating reactor vessel); GEA Exergy (supplier of flash dryer) were critically evaluated. The energy-saving potential is highest in recovering and reusing the latent heat of vaporization of water available in the exhaust of SSD. This is economically feasible if the dryer is coupled to a mechanical vapor compressor or if the dryer is integrated into a process using the steam. Further, as discussed in various literature works, the specific energy consumption for SSD is 0.45–0.88 against hot air drying, 2.9–4.1 MJ/kg of evaporated water.

Sobulska et al. [229] reviewed the superheated steam spray dryers (SSSD) and reported that, due to rapid moisture evaporation, particle porosity is very high improving the rehydration feature. Further, compared to conventional hot air dryer, SSSD will save 20%–30% energy due to change in medium and can be extended to 80% by appropriate utilization of exhaust steam including the surplus generated by moisture evaporation from feedstock.

The exhaust of the SSD contains low-temperature superheated steam used in drying and evaporated from the wet feedstock. The energy content is possible to be recovered and multiple options are available as discussed in detail by Mujumdar [9]. If efficient use of exhaust steam is carried out there would not be any emission from the dryer. Thus SSD utilization is safe for environment. Another aspect of sustainability is product quality, which is at par with available in the market or superior to that. In case of food products drying by superheated steam, the retention of nutrients, color, texture, etc. is significantly higher than the conventional hot air drying for majority of tested products. Overall, SSD is competent for sustainable development reducing significant energy consumption and emission and improves favorable product qualities.

## 8.3 PROSPECTS FOR SUPERHEATED STEAM DRYING

Future research in SSD needs to be interdisciplinary. Indeed, drying research has always been an interdisciplinary area with coupling of transport phenomenon and material science. Effective research in drying therefore requires expertise not only from engineering and physical sciences but also from economics, computer science, and mathematics. Figure 8.1 clearly shows the interdisciplinary nature of previous publications on SSD, which is drawn from the Scopus database (1929–2024, accessed on April 21, 2024) considering search keywords as "drying" OR "airless drying" OR "Superheated steam drying" and then refined with "Superheated steam drying" [32]. In future, there will be additional contribution from artificial intelligence as well. With close interaction between public and private enterprises there is strong potential for the development of new SSD technologies for diverse industrial applications of large scale.

**TABLE 8.1**
**Summary of Energy and Sustainability Aspect Studied in Literature of SSD**

| S.N. | Research Article | Dryer Type/Study | Sustainability Aspect | Key Findings/Remarks |
|---|---|---|---|---|
| 1. | Sobulska et al. [229] | Superheated steam spray drying, review from energy saving aspects-theoretical modeling, experimental, and pilot scale | Use of superheated steam for spray drying may reduce 20%–30% energy consumption, and further, it can be dropped to 80% by reuse of the exhaust steam. | • Initial work on spray drying with superheated steam is encouraging, showing good potential.<br>• More detailed study focusing on the physics and complex energy analysis of spray drying phenomena is lacking.<br>• Effect on thermophysical properties of various items of superheated steam is available for limited products. |
| 2. | Li et al. [29] | Review on energetic performances on superheated steam drying especially, rotary drum dryer, kiln dryer, impingement dryer, fluidized bed dryer, and, flash dryer. | Rotary drum and fluidized bed dryers with heat pipes in the drying chamber significantly improved energy utilization. | • Enhancing steam recycle reduces energy consumption in the impingement dryer; for example, savings in the total specific energy consumption can be up to 46% and 95% comparing with the hot air similar systems with and without exhaust air recycle, respectively. |
| 3. | Romdhana et al. [3] | Review on lab-scale and pilots SSD mainly on fluid bed, fixed bed, kiln, impingement jet, and flash drying.<br>Parameters: Energy efficiency | Due to lower energy consumption and better product quality, SSD will have high potential.<br>Due to the recovery of steam or integration or mechanical compression, the specific energy consumption is 4–9 times lower for various types of dryers with particular products compared to hot air dryers. | • Need to test a wide variety of products on pilot plants and should have available characteristics data to optimize the system.<br>• For SSD, there are certain challenges, such as loading and unloading of drying products, cost-effective and simple technology for cleaning steam, possible recirculation, process application for excess steam, etc.<br>• In addition to energy efficiency, other quality parameters need to be addressed. |

(*Continued*)

**TABLE 8.1 (*Continued*)**
**Summary of Energy and Sustainability Aspect Studied in Literature of SSD**

| S.N. | Research Article | Dryer Type/Study | Sustainability Aspect | Key Findings/Remarks |
|---|---|---|---|---|
| 4. | Mujumdar [9] | Critical review report on various SSD with a list of tried and probable products dried. | The probable energy saving due to the change from hot air to superheated steam will be around 30%.<br>Further excess steam is produced if having application elsewhere in the process, then the net energy consumption reduced by 4–5 times. | • Techno-economic analysis requires complete data for detailed evaluation, which was lacking.<br>• In addition to the energy savings, other important advantages such as improved product quality, reduction in pollution due to emissions and carbon dioxide, elimination of fire and explosion hazards, more efficient operation, more compact drying system, enhanced control, etc. need to be quantified to get the big picture of SSD. |
| 5. | Martynenko and Alves Vieira [234] | Proposed metrics for sustainability analysis of dryers based on system approach. | Sustainability assessment of drying is based on considering energy, exergy, environmental and economic aspects.<br>Metrics based on energy analysis: specific energy consumption and exergy analysis specific exergy consumption will give more insight for the comparison of different drying processes. | • Environmental effect due to drying can be investigated by sustainability index and renewability index. |

(*Continued*)

**TABLE 8.1 (*Continued*)**
**Summary of Energy and Sustainability Aspect Studied in Literature of SSD**

| S.N. | Research Article | Dryer Type/Study | Sustainability Aspect | Key Findings/Remarks |
|---|---|---|---|---|
| 6. | Elustondo et al. [158] | Wood drying advanced research directions | More fundamental and new approaches for drying modeling. | • Artificial Intelligence-based drying modeling needs a vast database of different wood species and drying behavior, which need to be developed. |
| 7. | Kumar et al. [235] | Tea drying overview for advanced drying and renewable energy as source | Reviewed SSD technology for tea sector along with others and renewable energy source based drying system. | • Work on SSD for tea leaves is very limited in literature and explored only at laboratory scaled models but showing positive results.<br>• A few studies on renewable energy mainly solar energy used either heat source along with direct heating as complimentary or only direct/indirect heating are performed in literature at lab-scale. More pilot and industrial-level studies are required to make them commercially viable. |

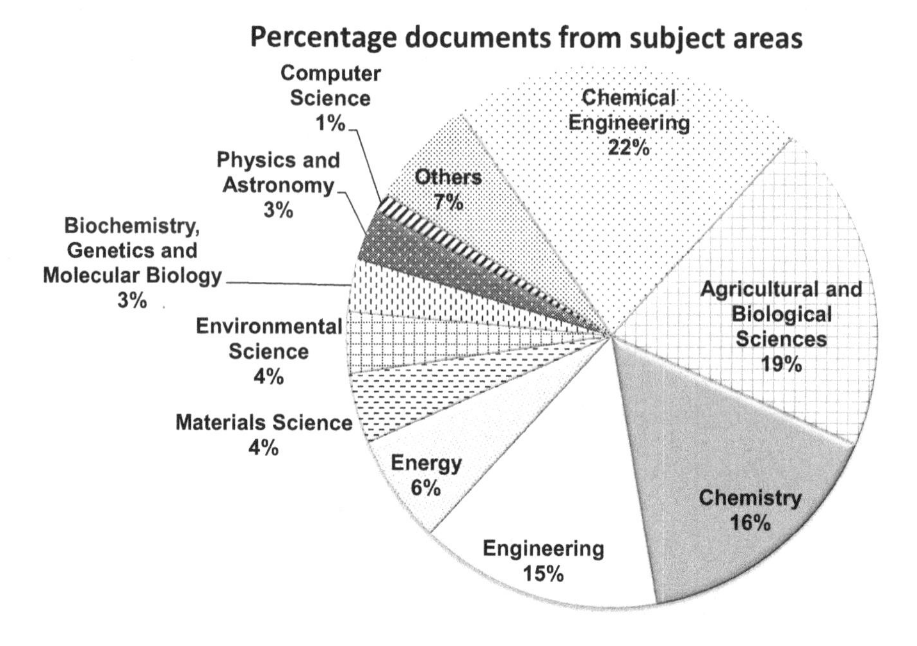


**FIGURE 8.1** Subject category-wise documents indexed in Scopus for SSD [32].

Currently, few dryers like fluidized bed, flash dryers, band dryers, impinging stream dryers, rotary dryers, and conveyor dryers working on superheated steam are commercially available in various industries. There are also lab-scale developments happening for spray dryers working on superheated steam for certain food products especially from dairy industry having significant potential. For heat-sensitive products such as food, fruits and vegetables, etc., low-pressure SSD is preferred. However, the system becomes complex due to maintenance of low-pressure and requires larger-size equipment because of lower drying rate. Additional heating systems like microwave, radiation, or conduction may be inserting heat pipe or jacketing will further increase the cost and becoming bulky. Largely, low-pressure SSD system becomes costly as well as complex for operation and control.

To get the advantages of superheated steam drying and reduce the adverse effects such as burning or others, hybrid drying with SSD is proposed, which is definitely having encouraging results and hence potential applications. However, the equipment becomes more complex and both the capital and operating costs increase. Although no publicly available reports consider the technoeconomics of such hybrid dryer concepts, it is expected that such drying systems may not be common in the near future.

A typical example of paper drying using superheated steam is discussed here to make it clear that, even after huge technical benefits, cost-effective engineering solutions are necessary to get market acceptance. Douglas (1994) has critically evaluated and summarized the results of an extensive series of experimental studies on laboratory-scale SSD of paper made from different grades of pulp. In general, the results are highly in favor of SSD, which can yield very high thermal efficiency if the dryer exhaust steam is utilized in the paper mill by mechanical compression or by external heating. Papers made using mechanical pulp are more economical and eco-friendly than chemical pulp. Also, it is possible to achieve higher-strength paper due to the presence of lignin. Technoeconomic analysis based on the series of results has shown superheated steam paper drying to be better than conventional cylinder drying of paper. A hybrid concept using a superheated steam dryer section followed by a conventional conduction dryer has been proposed as a good potential option. It is noteworthy that no experimental data are availing the public domain on a real dryer where the paper sheet is moving at a very high speed through the dryer section. The engineering design problem for this case is extremely challenging as the moving sheet will necessarily pump in finite volumes of non-vintage air into the superheated steam dryer making economic recovery of energy from the exhaust steam all but impossible. Although in principle, a dryer can be designed to overcome this challenge, the capital costs added risks involved, making it difficult to market this innovative dryer concept. If the energy costs soar or the carbon tax burden becomes very high, the concept may be revived at a future date; this, in our opinion, is an unrealistic scenario in the next couple of decades.

Although the original ideas proposed for drying newsprint and tissue paper using only superheat steam as the drying medium has not been successful in industrial acceptance, a new project which has been initiated in the European Union (EU) in January 2024 uses a different approach to achieve extremely favorable energy efficiency and reduction in greenhouse gas emissions [230]. This (a 9 million euro, 3 years) project involves joint research and development by a number of companies in the paper industrial sector along with several universities in the EU with highly relevant expertise. The goal of the ambitious project is to develop innovative steam dryers at a pilot scale to augment the capacity of existing dryers for drying materials in the form of continuous webs (sheets) [230].

## 8.4 GLOBAL MARKET DEMAND FOR SSD

The global industrial dryer size reached around USD 5.5 billion in 2023 and may grow with a compounded annualized growth rate of 3.8%–5.9% during 2024–2030 according to various market analysts' reports [231–233]. With huge market volumes and significant growth rate, there is always scope for advanced technology to be introduced, but definitely with certain features and precautions. Currently, there is an important burning issue of reduction of carbon footprint and improvement of green footprint due to the adverse effect of global warming. Looking toward these aspects, SSD if provided with a market-ready solution then certainly it will address these issues and it will show great market potential after commercialization.

Despite its long history and major validated advantages as a technologically sound drying concept for numerous high-volume materials, SSD has yet to make major inroads in the industrial arena. Indeed, over two decades ago Mujumdar [9] made some rather optimistic predictions about the market potential of SSD. These, however, have not materialized to any discernible extent. For commodities like wood, pulp, beet root, coal, etc., this technology was viable on an industrial scale although the capital and maintenance costs were higher relative to conventional dryers for the same materials. The primary advantages of SSDs are higher quality products and lower net energy consumption due to higher drying rate in addition to recovery of the latent heat from dryer exhaust steam. In some cases avoidance of fire and explosion hazards was a major added bonus. Although the market penetration potential predicted by Mujumdar [9] now appears to be overly optimistic, there is still a good possibility more vendors will offer this technology for wide range of products, once the issue of cost-effective cleaning and recovery of heat from lower dentally superheated steam exhaust from drying chamber gets cost-effective practical solution. It is noteworthy that almost all types of convective hot air dryers can be converted to SSD albeit at higher capital costs. For food materials, there are added advantages of pasteurization. For materials like various sludges high-temperature superheated steam can destroy microbial contaminants as well for safe disposal in addition to no odor issue because of closed-cycle operation.

The jury is still out on the future of this important concept for convective dryers. Interestingly, there is adequate knowledge base already to design and scale up the drying system; the roadblock is currently a cost-effective technology to harness energy from the exhaust. It is hoped that the problem will be solved within the next decade or two and SSD technology will be mainstream in the future for numerous commodity products. As evident this has major implications for the reduction of greenhouse gases released by air dryers leading to worsening the climate change issue. SSD systems in hybrid format by coupling it with microwave, radio frequency, or radiative heating also have industrial potential but much effort is required to enhance the technoeconomics. From the perspective of the use of renewable energy sources, although not much effort appears to have been carried out yet, there is a scope for research and development on the use of solar, wind, and biomass energy in SSD systems.

## 8.5 CLOSURE

The SSD seems to be attractive, having significant and vital benefits; however, there is still a wait to get its full potential commercially available. Recovery of exhaust steam is the single most critical obstacle to the wider application of steam drying technologies. In the case of food materials, higher product quality is the driving force for SSD. Lower energy costs are not an attractive basis for the selection of SSD relative to air drying as yet. If high or unique quality can command higher maker prices to offset the higher capital costs of typical SSD, then one could consider this technology for selected products. For drying of diverse sludge, biomass, low-rank coal, etc., SSD provides an excellent option as higher temperatures can be utilized without fear of fire or explosion. For heat-sensitive materials like fruits, vegetables, silk cocoons, etc., low-pressure SSD has industrial potential as the product quality is typically higher if air entrainment in the dryer can be prevented. Additionally, large-scale experience with several materials is typically not available in the public domain.

There is a tendency to prefer conventional dryer technologies to avoid potential risks in adopting new untested technology. This has hampered the wide acceptance of SSD in the industry. Much more research and development are needed to make innovative ideas of cleaning and retaining dryer exhaust steam cost-effectively market-ready. A new project has been initiated in the EU in January 2024, costing a 9 million euro with a duration of 3 years. The goal of this ambitious project is to develop innovative steam dryers at a pilot scale to augment the capacity of existing dryers for drying materials in the form of continuous webs (sheets). If many more of these types of projects come in the near future, then definitely this important technology may get market-ready, addressing the challenges appropriately in a short duration of time.

# Bibliography


[1] Mujumdar, A. S., Ed.; *Handbook of Industrial Drying*, 4th ed.; CRC Press, Taylor & Francis Group: Boca Raton, FL, 2014. https://doi.org/10.1201/b17208.

[2] Law, C. L.; Chen, H. H. H.; Mujumdar, A. S. Food Technologies: Drying. *Encyclop. Food Saf.*, **2014**; *3*, 156–167. https://doi.org/10.1016/B978-0-12-378612-8.00268-7.

[3] Romdhana, H.; Bonazzi, C.; Esteban-Decloux, M. Superheated Steam Drying: An Overview of Pilot and Industrial Dryers with a Focus on Energy Efficiency. *Dry. Technol.*, **2015**, *33*(10), 1255–1274. https://doi.org/10.1080/07373937.2015.1025139.

[4] Hausbrand, E. *Drying by Means of Air and Steam: Explanations, Formula, and Tables for Use in Practice*, First Engl.; Scott Greenwood and Co.: London, **1901**.

[5] Brar, N. K.; Ramachandran, R. P.; Cenkowski, S.; Paliwal, J. Effect of Superheated Steam- and Hot Air-Assisted Processing on Functional and Nutritional Properties of Yellow Peas. *Food Bioprocess Technol.*, **2021**, *14*(9), 1684–1699. https://doi.org/10.1007/s11947-021-02668-1.

[6] Pakowski, Z.; Adamski, R. On Prediction of the Drying Rate in Superheated Steam Drying Process. *Dry. Technol.*, **2011**, *29*(13), 1492–1498. https://doi.org/10.1080/07373937.2011.576320.

[7] Patel, S. K.; Bade, M. H. Superheated Steam Drying and Its Applicability for Various Types of the Dryer: The State of Art. *Dry. Technol.*, **2020**, *39*(3), 284–305. https://doi.org/10.1080/07373937.2020.1847139.

[8] Karrer, J. Drying by Electrically Superheated Steam. *Engineering*, **1920**, *110*(2868), 821–822.

[9] Mujumdar, A. S. *Superheated Steam Drying: Principles, Practice and Potential for Use of Electricity*; Canadian Electrical Association: Ottawa, ON, **1990**.

[10] Walker, W. H.; Lewis, W. K.; McAdams, W. H.; Gilliland, E. R. *Principles of Chemical Engineering*, 3rd ed.; McGraw-Hill Book Co.: New York, **1937**.

[11] Chu, J. C.; Lane, A. M.; Conklin, D. Evaporation of Liquids into Their Superheated Vapors. *Ind. Eng. Chem.*, **1953**, *45*(7), 1586–1591. https://doi.org/10.1021/ie50523a059.

[12] Eisenmann, E. Mechanische Holztrocknung Mit Trockenzentrifugen. *Holz-Zentralblatt*, **1950**, *76*, 106.

[13] Wenzel, L.; White, R. R. Drying Granular Solids in Superheated Steam. *Eng. Process Dev. Ind. Eng. Chem.*, **1951**, *43*(8), 1829–1837. https://doi.org/10.1021/ie50500a043.

[14] Dungler, J. Method for Drying Fibrous Sheet Material. *CLAIMS*, **1952**, 2590849.

[15] Beane, W. J. *Dehydration of Alfalfa with Superheated Steam, Kansas State College of Agriculture and Applied Science*; Kansas State University: Manhattan, NY, **1953**.

[16] Thompson, E. T. Dry Cheaper with Superheated Vapor. *Chem. Eng.*, **1955**, *62*(9), 104–108.

[17] Lane, A. M.; Stern, S. Application of Superheated Vapor Atmospheres to Drying. *Mech. Eng.*, **1956**, *78*(5), 423–426.

[18] Kauman, W. G. Equilibrium Moisture Content Relations Drying Control in Superheated Steam Drying. *For. Prod. J.*, **1956**, *6*(9), 328–332.

[19] Chu, J. C.; Finelt, S.; Hoerrner, W.; Lin, M.-S. Drying with Superheated Steam-Air Mixtures. *Ind. Eng. Chem.*, **1959**, *51*(3), 275–280. https://doi.org/10.1021/ie51394a033.

[20] Malmquist, L.; Noack, D. Untersuchungen Über Die Trocknung Empfindlicher Laubhölzer in ReinemHeißdampf(Ungesättigter Wasserdampf) Bei Unterdruck. Holz Als Roh - Und Werkstoff. *Mcmillen*, **1960**, *18*(5), 171–180.

[21] Hann, R. A. *An Investigation of the Drying of Wood at Temperatures above 100 Degrees Centigrade*; North Carolina State University: Raleigh, NC, **1965**.
[22] Yoshida, T.; Hyodo, T. Evaporation of Water in Air, Humid Air, and Superheated Steam. *Ind. Eng. Chem. Process Des. Dev.*, **1970**, *9*(2), 207–214. https://doi.org/10.1021/I260034A008/ASSET/I260034A008.FP.PNG_V03.
[23] Mikhailov, Y. A. *Superheated Steam Drying*; Energis: Moscow, **1967** (in Russian).
[24] Potter, O. E.; Beeby, C. J.; Fernando, W. J. N.; Ho, P. Drying Brown Coal in Steam-Heated, Steam-Fluidized Beds. *Dry. Technol.*, **1983**, *2*(2), 219–234. https://doi.org/10.1080/07373938308959826.
[25] Furukawa, T.; Akao, T. Deodorization by Superheated Steam Drying. *Dry. Technol.*, **1983**, *2*(3), 407–418. https://doi.org/10.1080/07373938308959839.
[26] Devahastin, S.; Mujumdar, A. S. Superheated Steam Drying of Foods and Biomaterials. *Modern Dry. Technol.*, **2014**, *5*, 57–84. https://doi.org/10.1002/9783527631704.ch03.
[27] Beeby, C.; Potter, O. E. Steam Drying. In *4th International Drying Symposium(IDS),* Kyoto, Japan; Toei, R., Mujumdar, A. S., Eds.; Springer: Berlin, Heidelberg., **1984**; pp. 41–58. https://doi.org/https://doi.org/10.1007/978-3-662-21830-3_5.
[28] Mujumdar, A. S. Superheated Steam Drying. In *Handbook of Industrial Drying*, 4th ed.; CRC Press, Taylor & Francis Group: Boca Raton, FL, **2014**; pp. 421–432. https://doi.org/10.1201/b17208.
[29] Jie, L.; Qian-Chao, L.; Bennamoun, L. Superheated Steam Drying: Design Aspects, Energetic Performances, and Mathematical Modeling. *Renew. Sustain. Energy Rev.*, **2016**, *60*, 1562–1583. https://doi.org/10.1016/j.rser.2016.03.033.
[30] Jensen, A. S.; Larsen, K. The Development of Large Pressurized Fluid Bed Steam Dryers from Fundamental Research to Industrial Plants. *Dry. Technol.*, **2015**, *33*(13). https://doi.org/10.1080/07373937.2015.1064944.
[31] Akao, T.; Aonuma, T. Puffing Drying of Grains. *In Proceedings of the First International Symposium on Drying*; Science Press: Princeton, NJ, **1987**; pp. 117–121.
[32] Elsevier, B. V. Scopus database. https://www.scopus.com/term/analyzer.uri?sort=plf-f&src=s&sid=c8ac4b290f13e73daf92b94eb4f7a55c&sot=a&sdt=a&sl=124&s=%28TITLE-ABS-KEY%28%22drying%22%29+OR+TITLE-ABS-KEY%28%22airless+drying%22%29+OR+TITLE-ABS-KEY%28%22Superheated+steam+drying%22%29%29+AND+P (accessed April 21, 2024).
[33] Espacenet Patent search. https://worldwide.espacenet.com/patent/search?f=pd%3Ain%3D18810101-20241231&q=%22Superheated Steam Drying%22(accessed March 1, 2024).
[34] Taechapairoj, C.; Dhuchakallaya, I.; Soponronnarit, S.; Wetchacama, S.; Prachayawarakorn, S. Superheated Steam Fluidised Bed Paddy Drying. *J. Food Eng.*, **2003**, *58*(1), 67–73. https://doi.org/10.1016/S0260-8774(02)00335-7.
[35] Devahastin, S.; Suvarnakuta, P.; Soponronnarit, S.; Mujumdar, A. S. A Comparative Study of Low-Pressure Superheated Steam and Vacuum Drying of a Heat-Sensitive Material. *Dry. Technol.*, **2004**, *22*(8), 1845–1867.
[36] Iyota, H.; Nishimura, N.; Onuma, T.; Nomura, T. Drying of Sliced Raw Potatoes in Superheated Steam and Hot Air. *Dry. Technol.*, **2001**, *19*(7), 1411–1424. https://doi.org/10.1081/DRT-100105297.
[37] Pang, S.; Dakin, M. Drying Rate and Temperature Profile for Superheated Steam Vacuum Drying and Moist Air Drying of Softwood Lumber. *Dry. Technol.*, **1999**, *17*(6), 1135–1147. https://doi.org/10.1080/07373939908917599.
[38] Levy, A.; Borde, I. Pneumatic and Flash Drying. In *Handbook of Industrial Drying*, 4th ed.; CRC Press, Taylor & Francis Group: Boca Raton, FL, **2014**.
[39] Bond, J. F.; Crotogino, R. H.; van Heiningen, A. R. P.; Douglas, W. J. M. An Experimental Study of the Falling Rate Period of Superheated Steam Impingement Drying of Paper. *Dry. Technol.*, **1992**, *10*(4), 961–977. https://doi.org/10.1080/07373939208916490.

[40] Krokida, M.; Marinos-Kouris, D.; Mujumdar, A. S. Rotary Drying. In *Handbook of Industrial Drying*, 4th ed.; CRC Press, Taylor & Francis Group: Boca Raton, FL, **2020**.
[41] Linke, T.; Happe, J.; Kohlus, R. Laboratory-Scale Superheated Steam Spray Drying of Food and Dairy Products. *Dry. Technol.*, **2022**, *40*(8), 1703–1714. https://doi.org/10.1080/07373937.2020.1870127.
[42] Perry, R. H. *Chemical Engineers' Handbook*, 5th ed.; McGraw-Hill: Kogakusha, Tokyo, **1973**.
[43] Iyota, H.; Nomura, T.; Nishimura, N. A Reverse Process of Superheated Steam Drying from Condensation to Evaporation. *Heat Transf. Asian Res.*, **1999**, *28*(5). https://doi.org/10.1002/(sici)1523-1496(1999)28:5<352::aid-htj2>3.0.co;2–j.
[44] Johnson, P.; Cenkowski, S.; Paliwal, J. *Superheated Steam Drying Characteristics of Single Cylindrical Compacts Produced from Wet Distiller's Spent Grain*. Winnipeg, MB: The Canadian Society of Bioengineering, **2013**.
[45] Ramachandran, R. P.; Paliwal, J.; Cenkowski, S. Computational Modelling of Superheated Steam Drying of Compacted Distillers' Spent Grain Coated with Solubles. *Food Bioprod. Process.*, **2019**, *116*, 63–77. https://doi.org/10.1016/j.fbp.2019.04.011.
[46] Luikov, A. V. *Heat and Mass Transfer in Capillary-Porous Bodies*, First Engl.; Pergamon Press Ltd: Oxford, **1966**.
[47] Yoshida, T.; Hyodo, T. Superheated Vapor Speeds Drying of Foods. *Food Eng.*, **1966**, *38*, 86–87.
[48] Trommelen, A. M.; Crosby, E. J. Evaporation and Drying of Drops in Superheated Vapors. *AIChE J.*, **1970**, *16*(5), 857–867. https://doi.org/10.1002/aic.690160527.
[49] Toei, R.; Okazaki, M.; Kimura, M.; Kubota, K. Drying Characteristics of a Porous Solid in Superheated Steam Drying. *Chem. Eng.*, **1966**, *30*(10), 949–950. https://doi.org/10.1252/KAKORONBUNSHU1953.30.949.
[50] Strumiłło, C.; Jones, P. L.; Z, R. Y. Miscellaneous Topics in Industrial Drying. In *Industrial Drying Handbook*; Mujumdar, A. S., Ed.; CRC Press Taylor & Francis Group, FL, **2014**; pp. 1077–1100.
[51] Liu, Y.; Kansha, Y.; Ishizuka, M.; Fu, Q.; Tsutsumi, A. Experimental and Simulation Investigations on Self-Heat Recuperative Fluidized Bed Dryer for Biomass Drying with Superheated Steam. *Fuel Process. Technol.*, **2015**, *136*, 79–86. https://doi.org/10.1016/J.FUPROC.2014.10.005.
[52] Chojnacka, K.; Mikula, K.; Izydorczyk, G.; Skrzypczak, D.; Witek-Krowiak, A.; Moustakas, K.; Ludwig, W.; Kułażyński, M. Improvements in Drying Technologies - Efficient Solutions for Cleaner Production with Higher Energy Efficiency and Reduced Emission. *J. Clean. Prod.*, **2021**, *320*, 128706. https://doi.org/10.1016/J.JCLEPRO.2021.128706.
[53] Costa, V. A. F.; Silva Neto da, F. On the Rate of Evaporation of Water into a Stream of Dry Air, Humidified Air and Superheated Steam, and the Inversion Temperature. *Int. J. Heat Mass Transf.*, **2003**, *46*(19), 3717–3726. https://doi.org/10.1016/S0017-9310(03)00174-1.
[54] Elustondo, D. M.; Mujumdar, A. S.; Urbicain, M. J. Optimum Operating Conditions in Drying Foodstuffs with Superheated Steam. *Dry. Technol.*, **2002**, *20*(2), 381–402.
[55] Pronyk, C.; Cenkowski, S.; Muir, W. E. Drying Foodstuffs with Superheated Steam. *Dry. Technol.*, **2004**, *22*(5), 899–916. https://doi.org/10.1081/DRT-120038571.
[56] Tang, Z.; Cenkowski, S. Dehydration Dynamics of Potatoes in Superheated Steam and Hot Air. *Can. Biosyst. Eng.*, **2000**, *42*(1), 1–13.
[57] Soponronnarit, S.; Prachayawarakorn, S.; Rordprapat, W.; Nathakaranakule, A.; Tia, W. A Superheated-Steam Fluidized-Bed Dryer for Parboiled Rice: Testing of a Pilot Scale and Mathematical Model Development. *Dry. Technol.*, **2006**, *24*(11), 1457–1467.
[58] Pakowski, Z.; Druzdzel, A.; Drwiega, J. Validation of a Model of an Expanding Superheated Steam Flash Dryer for Cut Tobacco Based on Processing Data. *Dry. Technol.*, **2004**, *22*(1–2), 45–57.

[59] Lum, A. L. C. *Superheated Steam in Spray Drying for Particle Functionality Engineering*; Monash University Clayton: Clayton, VIC, **2018**.

[60] Moreira, R. G. Impingement Drying of Foods Using Hot Air and Superheated Steam. *J. Food Eng.*, **2001**, *49*(4), 291–295. https://doi.org/10.1016/S0260-8774(00)00225-9.

[61] Wathanyoo, R.; Nathakaranakule, A.; Tia, W.; Soponronnarit, S. Comparative Study of Fluidized Bed Paddy Drying Using Hot Air and Superheated Steam. *J. Food Eng.*, **2005**, *71*, 28–36. https://doi.org/https://doi.org/10.1016/j.jfoodeng.2004.10.014.

[62] Prachayawarakorn, S.; Prachayawasin, P.; Soponronnarit, S. Heating Process of Soybean Using Hot-Air and Superheated-Steam Fluidized-Bed Dryers. *Food Sci. Technol.*, **2006**, *39*(7), 770–778. https://doi.org/10.1016/j.lwt.2005.05.013.

[63] Stokie, D.; Woo, M. W.; Bhattacharya, S. Comparison of Superheated Steam and Air Fluidized-Bed Drying Characteristics of Victorian Brown Coals. *Energy and Fuels*, **2013**, *27*(11), 6598–6606. https://doi.org/10.1021/ef401649j.

[64] Hanifzadeh, M.; Nabati, Z.; Longka, P.; Malakul, P.; Apul, D.; Kim, D.-S. Life Cycle Assessment of Superheated Steam Drying Technology as a Novel Cow Manure Management Method. *J. Environ. Manage.*, **2017**, *199*, 83–90. https://doi.org/10.1016/j.jenvman.2017.05.018.

[65] Jittanit, W.; Angkaew, K. Effect of Drying Schemes Using Superheated-Steam and Hot-Air as Drying Media on the Quality of Parboiled Chalky Rice Compared to Conventional Parboiling. *Dry. Technol.*, **2021**, *39*(16), 2218–2233. https://doi.org/10.1080/07373937.2020.1761375.

[66] Rahman, M. S. *Handbook of Food Preservation*, Vol. 3, 2nd ed.; M. Shafiur Rahman, Ed.; CRC Press Taylor & Francis Group: Boca Raton, FL, **2015**.

[67] Bourdoux, S.; Li, D.; Rajkovic, A.; Devlieghere, F.; Uyttendaele, M. Performance of Drying Technologies to Ensure Microbial Safety of Dried Fruits and Vegetables. *Compr. Rev. Food Sci. Food Saf.*, **2016**, *15*(6). https://doi.org/10.1111/1541-4337.12224.

[68] Alp, D.; Bulantekin, Ö. The Microbiological Quality of Various Foods Dried by Applying Different Drying Methods: A Review. *Eur. Food Res. Technol.*, **2021**, *247*(6), 1333–1343. https://doi.org/10.1007/S00217-021-03731-Z/TABLES/1.

[69] Jangam, S. V., Law, C. L., Mujumdar, A. S., Eds.; *Drying of Foods, Vegetables and Fruits*, Vol. 1, 1st edn.; Singapore, **2010**.

[70] Iyota, H.; Nishimura, N.; Yoshida, M.; Nomura, T. Simulation of Superheated Steam Drying Considering Initial Steam Condensation. *Dry. Technol.*, **2001**, *19*(7), 1425–1440. https://doi.org/10.1081/DRT-100105298.

[71] Namsanguan, Y.; Tia, W.; Devahastin, S.; Soponronnarit, S. Drying Kinetics and Quality of Shrimp Undergoing Different Two-Stage Drying Processes. *Dry. Technol.*, **2004**, *22*(4), 759–778. https://doi.org/10.1081/DRT-120034261.

[72] Leeratanarak, N.; Devahastin, S.; Chiewchan, N. Drying Kinetics and Quality of Potato Chips Undergoing Different Drying Techniques. *J. Food Eng.*, **2006**, *77*(3), 635–643. https://doi.org/10.1016/J.JFOODENG.2005.07.022.

[73] Alfy, A.; Kiran, B. V.; Jeevitha, G. C.; Hebbar, H. U. Recent Developments in Superheated Steam Processing of Foods: A Review. *Crit. Rev. Food Sci. Nutr.*, **2016**, *56*(13), 2191–2208. https://doi.org/10.1080/10408398.2012.740641.

[74] Iyota, H.; Konishi, Y.; Inoue, T.; Yoshida, K.; Nishimura, N.; Nomura, T. Popping of Amaranth Seeds in Hot Air and Superheated Steam. *Dry. Technol.*, **2005**, *23*(6), 1273–1287. https://doi.org/10.1081/DRT-200059502.

[75] Drying of Foods, *Vegetables and Fruits;* Vol. 2; Jangam, S. V., Law, C. L., Mujumdar, A. S., Eds.; Singapore, **2011**.

[76] Sablani, S. S. Drying of Fruits and Vegetables: Retention of Nutritional/Functional Quality. **2007**, *24*(2), 123–135. https://doi.org/10.1080/07373930600558904.

[77] Namsanguan, K.; Mangmool, P. Influence of Drying Conditions on the Drying of Longan without Stone by Low-Pressure Superheated. *Int. J. Mech. Prod. Eng.*, **2019**, *7*(9), 18–22.

[78] Lim, G. W.; Jafarzadeh, S.; Norazatul Hanim, M. R. Kinetic Study, Optimization and Comparison of Sun Drying and Superheated Steam Drying of Asam Gelugor(*Garcinia Cambogia*). *Food Res.*, **2020**, *4*(2). https://doi.org/10.26656/fr.2017.4(2).288.
[79] Husen, R. B. *Potential Use of Superheated-Steam Treatment in Underutilized Fruit of Engkala (Litsea Garciae) and Evaluation of Its Antioxidant Capacity, Graduate School of Life Science and Systems Engineering*; Japan: Kyushu Institute of Technology, **2015**.
[80] Jamradloedluk, J.; Nathakaranakule, A.; Soponronnarit, S.; Prachayawarakorn, S. Influences of Drying Medium and Temperature on Drying Kinetics and Quality Attributes of Durian Chip. *J. Food Eng.*, **2007**, *78*(1). https://doi.org/10.1016/j.jfoodeng.2005.09.017.
[81] Crank, J. *The Mathematics of Diffusion*, 2nd ed; Oxford: Clarendon Press, **1975**.
[82] Pimpaporn, P.; Devahastin, S.; Chiewchan, N. Effects of Combined Pretreatments on Drying Kinetics and Quality of Potato Chips Undergoing Low-Pressure Superheated Steam Drying. *J. Food Eng.*, **2007**, *81*(2). https://doi.org/10.1016/j.jfoodeng.2006.11.009.
[83] Cenkowski, S.; Pronyk, C.; Zmidzinska, D.; Muir, W. E. Decontamination of Food Products with Superheated Steam. *J. Food Eng.*, **2007**, *83*(1), 68–75. https://doi.org/10.1016/j.jfoodeng.2006.12.002.
[84] Suvarnakuta, P.; Devahastin, S.; Soponronnarit, S.; Mujumdar, A. S. Drying Kinetics and Inversion Temperature in a Low-Pressure Superheated Steam-Drying System. *Ind. Eng. Chem. Res.*, **2005**, *44*(6), 1934–1941. https://doi.org/10.1021/IE049675R.
[85] Suvarnakuta, P.; Devahastin, S.; Mujumdar, A. S. Drying Kinetics and β-Carotene Degradation in Carrot Undergoing Different Drying Processes. *J. Food Sci.*, **2005**, *70*(8), s520–s526. https://doi.org/10.1111/J.1365-2621.2005.TB11528.X.
[86] Pise, V. H.; Thorat, B. N. Green Steam for Sustainable Extraction of Essential Oils Using Solar Steam Generator: A Techno-Economic Approach. *Energy Nexus*, **2023**, *9*, 1–10. https://doi.org/https://doi.org/10.1016/j.nexus.2023.100175.
[87] Kozanoglu, B.; Vazquez, A. C.; Chanes, J. W.; Patiño, J. L. Drying of Seeds in a Superheated Steam Vacuum Fluidized Bed Coriender Pepper Seed. *J. Food Eng.*, **2006**, *75*, 383–387.
[88] Shaharuddin, S.; Husen, R.; Othman, A. Nutritional Values of Baccaurea Pubera and Comparative Evaluation of SHS Treatment on Its Antioxidant Properties. *J. Food Sci. Technol.*, **2021**, *58*(6), 2360–2367. https://doi.org/10.1007/S13197-020-04748-0.
[89] Malaikritsanachalee, P.; Choosri, W.; Choosri, T. Study on Intermittent Low-Pressure Superheated Steam Drying: Effect on Drying Kinetics and Quality Changes in Ripe Mangoes. *J. Food Process. Preserv.*, **2020**, *44*(9), 1–13. https://doi.org/10.1111/jfpp.14669.
[90] Eang, R.; Tippayawong, N. Optimization of Process Variables for Drying of Cashew Nuts by Superheated Steam. *Cogent Eng.*, **2018**, *5*(1). https://doi.org/10.1080/23311916.2018.1531457.
[91] Husen, R.; Andou, Y.; Ismail, A.; Shirai, Y.; Hassan, M. A.; Sciences, H.; Sciences, B.; Putra, U. Enhanced Polyphenol Content and Antioxidant Capacity in the Edible Portion of Avocado Dried with Superheated-Steam. *Int. J. Adv. Res.*, **2014**, *2*(8).
[92] Methakhup, S.; Chiewchan, N.; Devahastin, S. Effects of Drying Methods and Conditions on Drying Kinetics and Quality of Indian Gooseberry Flake. *Food Sci. Technol.*, **2005**, *38*(6), 579–587. https://doi.org/10.1016/J.LWT.2004.08.012.
[93] Chan, E. W. C.; Ong, A. C. L.; Lim, K. L.; Chong, W. Y.; Chia, P. X.; Foo, J. P. Y. Effects of Superheated Steam Drying on the Antioxidant and Anti-Tyrosinase Properties of Selected Labiatae Herbs. *Carpathian J. Food Sci. Technol.*, **2019**, *11*(1).
[94] Sehrawat, R.; Nema, P. K. Low Pressure Superheated Steam Drying of Onion Slices: Kinetics and Quality Comparison with Vacuum and Hot Air Drying in an Advanced Drying Unit. *J. Food Sci. Technol.*, **2018**, *55*(10), 4311–4320. https://doi.org/10.1007/s13197-018-3379-4.

[95] Kim, A.-N.; Ko, H.-S.; Kyo-Yean, L.; Rahman, M. S.; Heo, H. J.; Choi, S.-G. The Effect of Superheated Steam Drying on Physicochemical and Microbial Characteristics of Korean Traditional Actinidia (*Actinidia Arguta*) Leaves. *Korean Soc. Food Preserv.*, **2017**, *24*(3).
[96] Liu, J.; Xue, J.; Xu, Q.; Shi, Y.; Wu, L.; Li, Z. Drying Kinetics and Quality Attributes of White Radish in Low Pressure Superheated Steam. *Int. J. Food Eng.*, **2017**, *13*(7). https://doi.org/10.1515/ijfe-2016-0365.
[97] Phungamngoen, C.; Chiewchan, N.; Devahastin, S. Effects of Various Pretreatments and Drying Methods on Salmonella Resistance and Physical Properties of Cabbage. *J. Food Eng.*, **2013**, *115*(2), 237–244. https://doi.org/10.1016/J.JFOODENG.2012.10.020.
[98] Kingcam, R.; Devahastin, S.; Chiewchan, N. Effect of Starch Retrogradation on Texture of Potato Chips Produced by Low-Pressure Superheated Steam Drying. *J. Food Eng.*, **2008**, *89*(1), 72–79. https://doi.org/10.1016/J.JFOODENG.2008.04.008.
[99] Caixeta, A. T.; Moreira, R.; Castell-Perez, M. E. Impingement Drying of Potato Chips. *J. Food Process Eng.*, **2002**, *25*(1), 63–90. https://doi.org/10.1111/J.1745-4530.2002.TB00556.X.
[100] Van Deventer, H. C.; Heijmans, R. M. H. Drying with Superheated Steam. *Dry. Technol.*, **2001**, *19*(8), 2033–2045. https://doi.org/10.1081/DRT-100107287.
[101] Li, Y. B.; Seyed-Yagoobi, J.; Moreira, R.; Yamsaengsung, R. Superheated Steam Impingement Drying of Tortilla Chips. *Dry. Technol.*, **1999**, *17*(1–2), 191–213. https://doi.org/10.1080/07373939908917525.
[102] Bernardo, A. M. M.; Dumoulin, E. D.; Lebert, A. M.; Bimbenet, J.-J. Drying of Sugar Beet Fiber with Hot Air or Superheated Steam. *Dry. Technol.*, **1990**, *8*(4), 767–779. https://doi.org/10.1080/07373939008959914.
[103] Ma, K.; Ngamwonglumlert, L.; Devahastin, S.; Chindapan, N.; Chiewchan, N. Feasibility Study of the Use of Superheated Steam Spray Drying to Produce Selected Food Powders. *Dry. Technol.*, **2022**, *40*(12), 2445–2455. https://doi.org/10.1080/07373937.2021.1980886.
[104] Lee, K. Y.; Rahman, M. S.; Kim, A. N.; Jeong, E. J.; Kim, B. G.; Lee, M. H.; Kim, H. J.; Choi, S. G. Effect of Superheated Steam Treatment on Yield, Physicochemical Properties and Volatile Profiles of Perilla Seed Oil. *LWT*, **2021**, *135*. https://doi.org/10.1016/j.lwt.2020.110240.
[105] Choicharoen, K.; Devahastin, S.; Soponronnarit, S. Comparative Evaluation of Performance and Energy Consumption of Hot Air and Superheated Steam Impinging Stream Dryers for High-Moisture Particulate Materials. *Appl. Therm. Eng.*, **2011**, *31*(16), 3444–3452.
[106] Gómez, J. E.; Melo, D. L.; Bórquez, R. M.; Canales, E. R. Computational Study of Impingement Jet Drying of Seeds Using Superheated Steam Based on Kinetic Theory of Granular Flow. *Dry. Technol.*, **2009**, *27*(11), 1171–1182. https://doi.org/10.1080/07373930903262998.
[107] Yotaro, K.; Iyota, H.; Yoshida, K.; Moritani, J.; Inoue, T.; Nishimura, N.; Nomura, T. Effect of Moisture Content on the Expansion Volume of Popped Amaranth Seeds by Hot Air and Superheated Steam Using a Fluidized Bed System. *Biotechnol. Biochem.*, **2004**, *68*(10), 2186–2189. https://doi.org/10.1271/bbb.68.2186.
[108] Junka, N.; Rattanamechaiskul, C.; Wongs-Aree, C.; Soponronnarit, S. Drying and Mathematical Modelling for the Decelerated Rancidity of Treated Jasmine Brown Rice Using Different Drying Media. *J. Food Eng.*, **2021**, *289*, 110165. https://doi.org/10.1016/J.JFOODENG.2020.110165.
[109] Hampel, N.; Le, K. H.; Kharaghani, A.; Tsotsas, E. Continuous Modeling of Superheated Steam Drying of Single Rice Grains. *Dry. Technol.*, **2019**, *37*(12), 1583–1596.
[110] Chen, J.; Xu, Y. Analyzing the Characteristics of Roasting Process for Chinese Rice Wine by Fluidized Bed Using Superheated Steam. *Food Sci. Technol. Res.*, **2016**, *22*(2), 159–172. https://doi.org/10.3136/fstr.22.159.

[111] Kozanoglu, B.; Mazariegos, D.; Guerrero-Beltrán, J. A.; Welti-Chanes, J. Drying Kinetics of Paddy in a Reduced Pressure Superheated Steam Fluidized Bed. *Dry. Technol.*, **2013**, *31*(4), 452–461. https://doi.org/10.1080/07373937.2012.740543.

[112] Soponronnarit, S.; Prachayawarakorn, S.; Rordprapat, W.; Nathakaranakule, A.; Tia, W. A Superheated-Steam Fluidized-Bed Dryer for Parboiled Rice: Testing of a Pilot-Scale and Mathematical Model Development. *Dry. Technol.*, **2006**, *24*(11), 1457–1467. https://doi.org/10.1080/07373930600952800.

[113] Srivastav, S.; Kumbhar, B. Modeling Drying Kinetics of Paneer Using Artificial Neural Networks(ANN). *J. Food Res. Technol.*, **2014**, *2*(1), 39–45.

[114] Mayachiew, P.; Devahastin, S. Comparative Evaluation of Physical Properties of Edible Chitosan Films Prepared by Different Drying Methods. *Dry. Technol.*, **2008**, *26*(2). https://doi.org/10.1080/07373930701831309.

[115] Pronyk, C.; Cenkowski, S.; Muir, W. E.; Lukow, O. M. Effects of Superheated Steam Processing on the Textural and Physical Properties of Asian Noodles. *Dry. Technol.*, **2008**, *26*(2), 192–203. https://doi.org/10.1080/07373930701831382.

[116] Pronyk, C. *Effects of Superheated Steam Processing on the Drying Kinetics and Textural Properties of Instant Asian Noodles*; University of Manitoba: Winnipeg, MB, **2007**.

[117] Romdhana, H.; Bonazzi, C. Superheated Steam Drying of Apple Cubes in a Drying Kiln. In *Euro-Mediterranean Symposium for Fruit & Vegetable Processing;* Avignon: France, **2011**. https://doi.org/10.13140/2.1.2894.7845.

[118] Somjai, T.; Achariyaviriya, S.; Achariyaviriya, A.; Namsanguan, K. Strategy for Longan Drying in Two-Stage Superheated Steam and Hot Air. *J. Food Eng.*, **2009**, *95*(2), 313–321.

[119] Park, H. W.; Balasubramaniam, V. M.; Snyder, A. B.; Sekhar, J. A. Influence of Superheated Steam Temperature and Moisture Exchange on the Inactivation of Geobacillus Stearothermophilus Spores in Wheat Flour-Coated Surfaces. *Food Bioprocess Technol.*, **2022**, *15*(7), 1550–1562. https://doi.org/10.1007/s11947-022-02830-3.

[120] Devahastin, S.; Mujumdar, A. Applications for Fluidized Bed Drying. In *Handbook of Fluidization and Fluid-Particle Systems*; Yang, W.- C., Ed.; CRC Press Taylor & Francis Group: Boca Raton, FL, **2003**. https://doi.org/10.1201/9780203912744.ch18.

[121] Stokie, D. *A Study for the Future Application of Steam Fluidized Bed Drying of Victorian Brown Coal*; Monash University, Clayton: VIC, Australia, **2015**.

[122] *Steam Atmosphere Dryer Project: System Development and Field Test: Final Report*; OCLC/NY ID 42023098, **1999**.

[123] Jensen, A. S.; Morin, B. Energy and the Environment in Beet Sugar Production. *Sugar Ind.*, **2015**, 697–702. https://doi.org/10.36961/si16951.

[124] Kozanoglu, B.; Flores, A.; Guerrero-Beltrán, J. A.; Welti-Chanes, J. Drying of Pepper Seed Particles in a Superheated Steam Fluidized Bed Operating at Reduced Pressure. *Dry. Technol.*, **2012**, *30*(8). https://doi.org/10.1080/07373937.2012.675532.

[125] Chryat, Y.; Esteban-Decloux, M.; Labarde, C.; Romdhana, H. A Concept and Industrial Testing of a Superheated Steam Rotary Dryer Demonstrator: Cocurrent-Triple Pass Design. *Dry. Technol.*, **2019**, *37*(4), 468–474. https://doi.org/10.1080/07373937.2018.1460849.

[126] Xiao, H.; Mujumdar, A. S. Chapter 12: Impingement Drying: Applications and Future Trends. In *Drying Technologies for Foods: Fundamentals & Applications*; Nema, P. K., Kaur, B. P., Mujumdar, A. S. Eds.; CRC Press: Boca Raton, FL, 2018.

[127] Takahashi, N.; Satoh, Y.; Itoh, T.; Kojima, T.; Koya, T.; Furusawa, T.; Kunii, D. Drying of Slurry in Bed of Particles Fluidized by Pure Water Vapor. In *4th International Drying Symposium, DRYING'85;* Toei, R., Mujumdar, A. S., Eds.; Springer, Berlin: Kyoto, Japan, **1984**; pp 453–460.

[128] Potter, O. E.; Guang, L. X.; Georgakopoulos, S.; Ming, M. Q. Some Design Aspects of Steam-Fluidized Steam Heated Dryers. *Dry. Technol.*, **1990**, *8*(1), 25–39. https://doi.org/10.1080/07373939008959862.

[129] Faber, E. F. *Comparison between Air Drying and Steam Drying in a Fluidized Bed*; University of Natal (Current University of KwaZulu-Natal): South Africa, **1991**.

[130] Berghel, J.; Renström, R. Usefulness and Significance of Energy and Mass Balances of a Fluidized Superheated Steam Dryer. *Dry. Technol.*, **2001**, *19*(6), 1083–1098. https://doi.org/10.1081/DRT-100104806.

[131] Berghel, J.; Renström, R. Basic Design Criteria and Corresponding Results Performance of a Pilot-Scale Fluidized Superheated Atmospheric Condition Steam Dryer. *Biomass and Bioenergy*, **2002**, *23*(2), 103–112. https://doi.org/10.1016/S0961-9534(02)00040-5.

[132] Klutz, H.-J.; Moser, C.; Bargen, N. von. The RWE Power WTA Process (Fluidized Bed Drying) as a Key for Higher Efficiency. *Górnictwo i Geoinżynieria*, **2011**, *35*(2), 147–153.

[133] Shi, Y.; Xiao, Z.; Wang, Z.; Liu, X.; Yang, D. Numerical Simulation on Superheated Steam Fluidized Bed Drying: II. Experiments and Numerical Simulation. *Dry. Technol.*, **2011**, *29*(11). https://doi.org/10.1080/07373937.2011.592050.

[134] Aziz, M.; Oda, T.; Kashiwagi, T. Innovative Steam Drying of Empty Fruit Bunch with High Energy Efficiency. *Dry. Technol.*, **2015**, *33*(4), 395–405. https://doi.org/10.1080/07373937.2014.970257.

[135] Hulkkonen, S.; Heinonen, O.; Tiihonen, J.; Impola, R. Drying of Wood Biomass at High Pressure Steam Atmosphere; Experimental Research and Application. *Dry. Technol.*, **1994**, *12*(4). https://doi.org/10.1080/07373939408959999.

[136] Fyhr, C. *Steam Drying of Wood Chips in Pneumatic Conveying Dryers*; Chalmers University of Technology: Gothenburg, Sweden, **1996**.

[137] Blasco, R.; Alvarez, P. I. Flash Drying of Fish Meals with Superheated Steam: Isothermal Process. *Dry. Technol.*, **1999**, *17*(4–5). https://doi.org/10.1080/07373939908917569.

[138] Blasco, R.; Vega, R.; Alvarez, P. I. Pneumatic Drying with Superheated Steam: Bi-Dimensional Model for High Solid Concentration. *Dry. Technol.*, **2001**, *19*(8). https://doi.org/10.1081/DRT-100107288.

[139] Douglas, W. J. M. Drying Paper in Superheated Steam. *Dry. Technol.*, **1994**, *12*(6). https://doi.org/10.1080/07373939408961009.

[140] Shiravi, A. H. *Drum Drying of Black Liquor Using Superheated Steam Impinging Jets*; McGill University, Montreal, QC, Canada, **1995**.

[141] Shiravi, A. H.; Mujumdar, A. S.; Kubes, G. J. Drum Drying of Black Liquor Using Superheated Steam Impinging Jets. *Dry. Technol.*, **1997**, *15*(5), 1571–1584. https://doi.org/10.1080/07373939708917308.

[142] Bonazzi, C.; Dumoulin, E.; Raoult-Wack, A.-L.; Berk, Z.; Bimbenet, J. J.; Courtois, F.; Trystram, G.; Vasseur, J. Food Drying and Dewatering. *Dry. Technol.*, **1996**, *14*(9), 2135–2170. https://doi.org/10.1080/07373939608917199.

[143] Borquez, R. M.; Canales, E. R.; Quezada, H. R. Drying of Fish Press-Cake with Superheated Steam in a Pilot Plant Impingement System. *Dry. Technol.*, **2008**, *26*(3). https://doi.org/10.1080/07373930801897986.

[144] Fuengfoo, M.; Devahastin, S.; Niumnuy, C.; Soponronnarit, S. Preliminary Study of Superheated Steam Spray Drying: A Case Study with Maltodextrin; **2019**. https://doi.org/10.4995/ids2018.2018.7881.

[145] Swasdisevi, T.; Devahastin, S.; Thanasookprasert, S.; Soponronnarit, S. Comparative Evaluation of Hot-Air and Superheated-Steam Impinging Stream Drying as Novel Alternatives for Paddy Drying. *Dry. Technol.*, **2013**, *31*(6). https://doi.org/10.1080/07373937.2013.773908.

[146] Islam, M. Z.; Kitamura, Y.; Yamano, Y.; Kitamura, M. Effect of Vacuum Spray Drying on the Physicochemical Properties, Water Sorption and Glass Transition Phenomenon of Orange Juice Powder. *J. Food Eng.*, **2016**, *169*. https://doi.org/10.1016/j.jfoodeng.2015.08.024.

[147] Lum, A.; Cardamone, N.; Beliavski, R.; Mansouri, S.; Hapgood, K.; Woo, M. W. Unusual Drying Behaviour of Droplets Containing Organic and Inorganic Solutes in Superheated Steam. *J. Food Eng.*, **2019**, *244*. https://doi.org/10.1016/j.jfoodeng.2018.09.021.

[148] Lum, A.; Mansouri, S.; Hapgood, K.; Woo, M. W. Single Droplet Drying of Milk in Air and Superheated Steam: Particle Formation and Wettability. *Dry. Technol.*, **2018**, *36*(15). https://doi.org/10.1080/07373937.2017.1416626.

[149] Lum, A.; Cardamone, N.; Beliavski, R.; Mansouri, S.; Hapgood, K.; Woo, M. W. Role of Steam as a Medium for Droplet Crystallization. *Ind. Eng. Chem. Res.*, **2019**, *58*(19). https://doi.org/10.1021/acs.iecr.9b00561.

[150] Kanani, K.; Gupta, A. K.; Patel, S. K.; Bade, M. H. Exploration of Climate Zones Based on Hierarchal Clustering Algorithm for Buildings in India. *J. Build. Pathol. Rehabil.*, **2022**, *7*(1), 1–12. https://doi.org/10.1007/S41024-022-00210-0/METRICS.

[151] Chiewchan, N.; Pakdee, W.; Devahastin, S. Effect of Water Activity on Thermal Resistance of Salmonella Krefeld in Liquid Medium and on Rawhide Surface. *Int. J. Food Microbiol.*, **2007**, *114*(1). https://doi.org/10.1016/j.ijfoodmicro.2006.10.037.

[152] Goula, A. M.; Adamopoulos, K. G. A New Technique for Spray Drying Orange Juice Concentrate. *Innov. Food Sci. Emerg. Technol.*, **2010**, *11*(2), 342–351. https://doi.org/10.1016/j.ifset.2009.12.001.

[153] Shrestha, A. K.; Ua-Arak, T.; Adhikari, B. P.; Howes, T.; Bhandari, B. R. Glass Transition Behavior of Spray Dried Orange Juice Powder Measured by Differential Scanning Calorimetry(DSC) and Thermal Mechanical Compression Test(TMCT). *Int. J. Food Prop.*, **2007**, *10*(3). https://doi.org/10.1080/10942910601109218.

[154] Roos, Y. H.; Finley, J. W.; DeMan, J. M. Water. In *Principles of Food Chemistry: Food Science Text Series*; deMan, J. M., , Finley, J. W., Hurst, W. J., Lee, C. Y., Eds.; Springer, Cham, **2018**; pp 1–37. https://doi.org/https://doi.org/10.1007/978-3-319-63607-8_1.

[155] Roos, Y. H. Glass Transition Temperature and Its Relevance in Food Processing. *Annu. Rev. Food Sci. Technol.*, **2010**, *1*(1). https://doi.org/10.1146/annurev.food.102308.124139.

[156] Pise, V. H.; Harlalka, R.; Thorat, B. N. *Drying of Aromatic Plant Material for Natural Perfumes*, 1st edn; CRC Press Taylor & Francis Group: Boca Raton, FL, **2023**. https://doi.org/10.1201/9781003315384.

[157] Roya, S. *Drying of Hog Fuel in a Fixed Bed*; The University of British Columbia, **1990**.

[158] Elustondo, D.; Matan, N.; Langrish, T.; Pang, S. Advances in Wood Drying Research and Development. *Dry. Technol.*, **2023**, *41*(6). https://doi.org/10.1080/07373937.2023.2205530.

[159] Chen, Z. *Primary Driving Force in Wood Vacuum Drying*, Faculty of the Virginia Polytechnic Institute and State University, **1997**.

[160] Espinoza, O.; Bond, B. Vacuum Drying of Wood-State of the Art. *Curr. For. Reports*, **2016**, *2*(4), 223–235. https://doi.org/10.1007/S40725-016-0045-9.

[161] Shibata, H.; Mada, J.; Shinohara, H. Steam Drying of Sintered Glass Bead Spheres under Vacuum. *Ind. Eng. Chem. Res.*, **1988**, *27*(12). https://doi.org/10.1021/ie00084a024.

[162] Mada, J.; Shibata, H.; Funatsu, K. Prediction of Drying Rate Curves on Sintered Spheres of Glass Beads in Superheated Steam under Vacuum. *Ind. Eng. Chem. Res.*, **1990**, *29*(4), 614–617. https://doi.org/10.1021/ie00100a018.

[163] Nimmol, C.; Devahastin, S.; Swasdisevi, T.; Soponronnarit, S. Drying of Banana Slices Using Combined Low-Pressure Superheated Steam and Far-Infrared Radiation. *J. Food Eng.*, **2007**, *81*(3), 624–633. https://doi.org/10.1016/j.jfoodeng.2006.12.022.

[164] Thomkapanich, O.; Suvarnakuta, P.; Devahastin, S. Study of Intermittent Low-Pressure Superheated Steam and Vacuum Drying of a Heat-Sensitive Material. *Dry. Technol.*, **2007**, *25*(1). https://doi.org/10.1080/07373930601161146.

[165] Tang, H. Y. Design and Functioning of Low Pressure Superheated Steam *Processing Unit*; University of Manitoba" Winnipeg, MB, **2010**.

[166] Messai, S.; Sghaier, J.; El Ganaoui, M.; Chrusciel, L.; Gabsi, S. Low-Pressure Superheated Steam Drying of a Porous Media. *Dry. Technol.*, **2015**, *33*(1), 103–110. https://doi.org/10.1080/07373937.2014.933843.

[167] Messai, S.; Sghaier, J.; Chrusciel, L.; El Ganaoui, M.; Gabsi, S. Low-Pressure Superheated Steam Drying-Vacuum Drying of a Porous Media and the Inversion Temperature. *Dry. Technol.*, **2015**, *33*(1). https://doi.org/10.1080/07373937.2014.933844.

[168] Narmatha, I.; Ganapathy, S.; Balakrishnan, M.; Geethalakshmi, I.; Subramanian, P. Superheated Steam Drying of Potato Slices under Low Pressure Conditions. *Biol. Forum An Int. J.*, **2022**, *14*(1), 919–925.

[169] Pang, S.; Pearson, H. Experimental Investigation and Practical Application of Superheated Steam Drying Technology for Softwood Timber. *Dry. Technol.*, **2004**, *22*(9), 2079–2094. https://doi.org/10.1081/DRT-200034252.

[170] Yamsaengsung, R.; Sattho, T. Superheated Steam Vacuum Drying of Rubberwood. *Dry. Technol.*, **2008**, *26*(6). https://doi.org/10.1080/07373930802046518.

[171] Redman, A. *Evaluation of Super-Heated Steam Vacuum Drying Viability and Development of a Predictive Drying Model for Four Australian Hardwood Species*; PROJECT NUMBER: PNB045-0809, FWPA, 2011.

[172] Elustondo, D.; Ahmed, S.; Oliveira, L. Drying Western Red Cedar with Superheated Steam. *Dry. Technol.*, **2014**, *32*(5). https://doi.org/10.1080/07373937.2013.843190.

[173] He, Z.; Qiu, S.; Zhang, Y.; Zhao, Z.; Yi, S. Heat Transfer Characteristics during Superheated Steam Vacuum Drying of Poplar. *Forest Prod. J.* **2016**, *66* (5–6), 308–312. https://doi.org/10.13073/FPJ-D-15-00054.

[174] Tolstorebrov, I.; Bantle, M.; Hafner, A.; Kuz, B.; Eikevik, T. M. Energy Efficiency by Vapor Compression in Super Heated Steam Drying Systems. In *11th IIR Gustav Lorentzen Conference on Natural Refrigerants: Natural Refrigerants and Environmental Protection, GL 2014;* IIF-IIR, **2014**.

[175] Atsonios, K.; Violidakis, I.; Agraniotis, M.; Grammelis, P.; Nikolopoulos, N.; Kakaras, E. Thermodynamic Analysis and Comparison of Retrofitting Pre-Drying Concepts at Existing Lignite Power Plants. *Appl. Therm. Eng.*, **2015**, *74*. https://doi.org/10.1016/j.applthermaleng.2013.11.007.

[176] Liu, M., Xu, C.; Han, X.; Liu, R.; Qin, Y.; Yan, J. Integration of Evaporative Dryers into Lignite-Fired Power Plants: A Review. *Dry. Technol.*, **2020**, *38*(15), 1996–2014. https://doi.org/10.1080/07373937.2019.1606824.

[177] Chantasiriwan, S. The Use of Steam Dryer with Heat Recovery to Decrease the Minimum Exhaust Flue Gas Temperature and Increase the Net Efficiency of Thermal Power Plant. *Dry. Technol.*, **2023**, *41*(5). https://doi.org/10.1080/07373937.2022.2113406.

[178] Dibella, F. A.; Doyle, E. F.; Becker, F. E.; Lang, R. *Steam Atmosphere Drying Concepts Using Steam Exhaust Recompression*; USDE: Idaho Falls, ID, **1991**.

[179] Shi-Ruo, C.; Jin-Yong, C.; Mujumdar, A. S. A Preliminary Study of Steam Drying of Silkworm Cocoons. *Drying Technol.*, **1992**. https://doi.org/10.1080/07373939208916425.

[180] Wimmerstedt, R. Steam Drying - History and Future. *Dry. Technol.*, **1995**, *13*(5–7). https://doi.org/10.1080/07373939508917009.

[181] Fitzpatrick, J. Sludge Processing by Anaerobic Digestion and Superheated Steam Drying. *Water Res.*, **1998**, *32*(10), 2897–2902.

[182] Aly, S. E. Energy Efficient Combined Superheated Steam Dryer/MED. *Appl. Therm. Eng.*, **1999**, *19*(6), 659–668. https://doi.org/10.1016/S1359-4311(98)00073-8.

[183] Alves-Filho, O.; Roos, Y. H. Advances in Multi-Purpose Drying Operations with Phase and State Transitions. *Dry. Technol.*, **2006**, *24*(3). https://doi.org/10.1080/07373930600564357.

[184] R Vance Morey; Huixiao Zheng; Matthew V Pham; Nalladurai Kaliyan. Superheated Steam Drying Technology in an Ethanol Production Process. In *Paper Number: 1009069, ASABE Annual International Meeting, ASABE Annual International Meeting; American Society of Agricultural and Biological Engineers: St.* Joseph, MI, **2010**. https://doi.org/10.13031/2013.32713.

[185] Morey, R. V.; Zheng, H.; Kaliyan, N.; Pham, M. V. Modelling of Superheated Steam Drying for Combined Heat and Power at a Corn Ethanol Plant Using Aspen Plus Software. *Biosyst. Eng.*, **2014**, *119*, 80–88. https://doi.org/10.1016/j.biosystemseng.2014.02.001.
[186] Aziz, M.; Kansha, Y.; Tsutsumi, A. Self-Heat Recuperative Fluidized Bed Drying of Brown Coal. *Chem. Eng. Process. Process Intensif.*, **2011**, *50*(9). https://doi.org/10.1016/j.cep.2011.07.005.
[187] Aziz, M.; Kansha, Y.; Kishimoto, A.; Kotani, Y.; Liu, Y.; Tsutsumi, A. Advanced Energy Saving in Low Rank Coal Drying Based on Self-Heat Recuperation Technology. *Fuel Process. Technol.*, **2012**, *104*. https://doi.org/10.1016/j.fuproc.2012.06.020.
[188] Aziz, M.; Oda, T.; Kashiwagi, T. Enhanced High Energy Efficient Steam Drying of Algae. Appl. Energy, **2013**, *109*, 163–170. https://doi.org/10.1016/j.apenergy.2013.04.004.
[189] Aziz, M.; Prawisudha, P.; Prabowo, B.; Budiman, B. A. Integration of Energy-Efficient Empty Fruit Bunch Drying with Gasification/Combined Cycle Systems. Appl. Energy, **2015**, *139*. https://doi.org/10.1016/j.apenergy.2014.11.038.
[190] Yamsaengsung, R.; Tabtiang, S. Hybrid Drying of Rubberwood Using Superheated Steam and Hot Air in a Pilot-Scale. *Dry. Technol.*, **2011**, *29*(10), 1170–1178. https://doi.org/10.1080/07373937.2011.574805.
[191] Ratnasingam, J.; Grohmann, R. Superheated Steam Application to Optimize the Kiln Drying of Rubberwood (Hevea Brasiliensis). *Eur. J. Wood Wood Prod.*, **2015**, *73*(3). https://doi.org/10.1007/s00107-015-0898-9.
[192] Johnson, P. *Energy and Exergy Analysis in Spray Drying Systems*, The University of Sydney, **2019**.
[193] Jaszczur, M.; Dudek, M.; Rosen, M. A.; Kolenda, Z. An Analysis of Integration of a Power Plant with a Lignite Superheated Steam Drying Unit. *J. Clean. Prod.*, **2020**, *243*. https://doi.org/10.1016/j.jclepro.2019.118635.
[194] Chantasiriwan, S.; Charoenvai, S. Improving the Performance of Cogeneration System in Sugar Factory by the Integration of Superheated Steam Dryer and Parabolic Trough Solar Collector; In: *Renewable Energy and Sustainable Buildings. Innovative Renewable Energy*; Sayigh, A., Eds.; Springer, Cham, **2020**. https://doi.org/10.1007/978-3-030-18488-9_75.
[195] Yao, Y.; Lu, Z.; Gong, Y.; Guo, S.; Xiao, C.; Hu, W. Drying Characteristics of a Combined Drying System of Low-Pressure Superheated Steam and Heat Pump. *Process.*, **2022**, *10*(7), 1402. https://doi.org/10.3390/PR10071402.
[196] Hager, J.; Hermansson, M.; Wimmerstedt, R. Modelling Steam Drying of a Single Porous Ceramic Sphere: Experiments and Simulations. Chem. Eng. Sci., **1997**, *52*(8), 1253–1264. https://doi.org/10.1016/S0009-2509(96)00493-9.
[197] Chen, Z.; Wu, W.; Agarwal, P. K. Steam-Drying of Coal. Part 1. Modeling the Behavior of a Single Particle. Fuel, **2000**, *79*(8), 961–974. https://doi.org/10.1016/S0016-2361(99)00217-3.
[198] Hosseinalipour, S. M.; Mujumdar, A. S. Superheated Steam Drying of a Single Particle in an Impinging Stream Dryer. Dry. Technol., **1995**, *13*(5–7). https://doi.org/10.1080/07373939508917022.
[199] Kiriyama, T.; Sasaki, H.; Hashimoto, A.; Kaneko, S.; Maeda, M. Experimental Observations and Numerical Modeling of a Single Coarse Lignite Particle Dried in Superheated Steam. Mater. *Trans.*, **2013**, *54*(9). https://doi.org/10.2320/matertrans.M-M2013817.
[200] Hao, X.; Yu, C.; Zhang, G.; Li, X.; Wu, Y.; Lv, J. Modeling Moisture and Heat Transfer during Superheated Steam Wood Drying Considering Potential Evaporation Interface Migration. *Dry. Technol.*, **2020**, *38*(15). https://doi.org/10.1080/07373937.2019.1662801.
[201] Le, K. H.; Thu, T.; Tran, H.; Nguyen, N. A.; Kharaghani, A. Multiscale Modeling of Superheated Steam Drying of Particulate Materials. **2020**, *5*, 913–922. https://doi.org/10.1002/ceat.201900602.
[202] Le, K. H.; Tsotsas, E.; Kharaghani, A. Continuum-Scale Modeling of Superheated Steam Drying of Cellular Plant Porous Media. *Int. J. Heat Mass Transf.*, **2018**, *124*, 1033–1044. https://doi.org/10.1016/j.ijheatmasstransfer.2018.04.032.

[203] Le, K. H.; Tran, T. T. H.; Tsotsas, E.; Kharaghani, A. Superheated Steam Drying of Single Wood Particles: Modeling and Comparative Study with Hot Air Drying. *Chem. Eng. Technol.*, **2021**, *44*(1). https://doi.org/10.1002/ceat.202000133.

[204] Olufemi, B. A.; Udefiagbon, I. F. Modelling the Drying of Porous Coal Particles in Superheated Steam. *Chem. Biochem. Eng. Q.*, **2010**, *24*(1).

[205] Heinrich, S.; Krüger, G.; Mörl, L. Modelling of the Batch Treatment of Wet Granular Solids with Superheated Steam in Fluidized Beds. Chem. Eng. Process., **1999**, *38*(2), 131–142. https://doi.org/10.1016/S0255-2701(98)00082-8.

[206] Taechapairoj, C.; Prachayawarakorn, S.; Soponronnarit, S. Modelling of Parboiled Rice in Superheated-Steam Fluidized Bed. *J. Food Eng.*, **2006**, *76*(3). https://doi.org/10.1016/j.jfoodeng.2005.05.040.

[207] Kovenskii, V. I.; Borodulya, V. A.; Teplitskii, Y. S.; Pal'Chenok, G. I.; Slizhuk, D. S. Modeling of Superheated-Steam Drying of Biofuel in a Fluidized Bed. *J. Eng. Phys. Thermophys.*, **2010**, *83*(4), 764–769. https://doi.org/10.1007/S10891-010-0395-2.

[208] Chen, Z.; Agarwal, P. K.; Agnew, J. B. Steam Drying of Coal. Part 2. Modeling the Operation of a Fluidized Bed Drying Unit. *Fuel*, **2001**, *80*, 209–223. https://doi.org/10.1016/S0016-2361(00)00081-8.

[209] Tang, Z.; Cenkowski, S.; Muir, W. E. Modelling the Superheated-Steam Drying of a Fixed Bed of Brewers' Spent Grain. *Biosyst. Eng.*, **2004**, *87*(1). https://doi.org/10.1016/j.biosystemseng.2003.09.008.

[210] Tran, T. T. H. Modelling of Drying in Packed Bed by Super Heated Steam. *J. Mech. Eng. Res. Dev.*, **2020**, *43*(1), 135.

[211] Messai, S.; Sghaier, J.; Belghith, A. Mathematical Modeling of a Packed Bed Drying with Humid Air and Superheated Steam. *J. Porous Media*, **2011**, *14*(2). https://doi.org/10.1615/JPorMedia.v14.i2.50.

[212] Sghaier, J.; Messai, S.; Jomaa, W.; Belghith, A. Modeling Heat and Mass Transfer during Superheated Steam Drying of a Fixed Bed of Porous Particles. *J. Porous Media*, **2009**, *12*(7). https://doi.org/10.1615/JPorMedia.v12.i7.30.

[213] Malekjani, N.; Jafari, S. M. Simulation of Food Drying Processes by Computational Fluid Dynamics (CFD); Recent Advances and Approaches. *Trends Food Sci. Technol.*, **2018**, *78*(Complete), 206–223. https://doi.org/10.1016/j.tifs.2018.06.006.

[214] Ramachandran, R. P.; Akbarzadeh, M.; Paliwal, J.; Cenkowski, S. Three-Dimensional CFD Modelling of Superheated Steam Drying of a Single Distillers' Spent Grain Pellet. *J. Food Eng.*, **2017**, *212*, 121–135. https://doi.org/10.1016/j.jfoodeng.2017.05.025.

[215] Kimwa, M. J.; Karuri, N.; Tanui, J. Computational Modeling of Spatial Variation in Moisture Content and Temperature Distribution in Corn at Different Superheated Steam Temperatures. *Cogent Eng.*, **2023**, *10*(1). https://doi.org/10.1080/23311916.2023.2216864.

[216] Frydman, A.; Vasseur, J.; Moureh, J.; Sionneau, M.; Tharrault, P. Comparison of Superheated Steam and Air Operated Spray Dryers Using Computational Fluid Dynamics. *Dry. Technol.*, **1998**, *16*(7). https://doi.org/10.1080/07373939808917464.

[217] Xiao, Z. F.; Zhang, F.; Wu, N. X.; Liu, X. D. CFD Modeling and Simulation of Superheated Steam Fluidized Bed Drying Process. *IFIP Adv. Inf. Commun. Technol.*, **2013**, *392*. https://doi.org/10.1007/978-3-642-36124-1_18.

[218] Mohseni, M.; Kolomijtschuk, A.; Peters, B.; Demoulling, M. Biomass Drying in a Vibrating Fluidized Bed Dryer with a Lagrangian-Eulerian Approach. *Int. J. Therm. Sci.*, **2019**, *138*. https://doi.org/10.1016/j.ijthermalsci.2018.12.038.

[219] Frydman, A.; Vasseur, J.; Ducept, F.; Sionneau, M.; Moureh, J. Simulation of Spray Drying in Superheated Steam Using Computational Fluid Dynamics. *Dry. Technol.*, **1999**, *17*(7–8), 1313–1326. https://doi.org/10.1080/07373939908917617.

[220] Suvarnakuta, P.; Devahastin, S.; Mujumdar, A. S. A Mathematical Model for Low-Pressure Superheated Steam Drying of a Biomaterial. *Chem. Eng. Process. Process Intensif.*, **2007**, *46*(7), 675–683. https://doi.org/10.1016/j.cep.2006.09.002.

[221] Kittiworrawatt, S.; Devahastin, S. Improvement of a Mathematical Model for Low-Pressure Superheated Steam Drying of a Biomaterial. *Chem. Eng. Sci.*, **2009**, *64*(11), 2644–2650. https://doi.org/10.1016/J.CES.2009.02.036.

[222] Elustondo, D.; Elustondo, M. P.; Urbicain, M. J. Mathematical Modeling of Moisture Evaporation from Foodstuffs Exposed to Subatmospheric Pressure Superheated Steam. *J. Food Eng.*, **2001**, *49*(1), 15–24. https://doi.org/10.1016/S0260-8774(00)00180-1.

[223] Pang, S. Some Considerations in Simulation of Superheated Steam Drying of Softwood Lumber. *Dry. Technol.*, **1997**, *15*(2). https://doi.org/10.1080/07373939708917252.

[224] Adamski, R.; Siuta, D.; Kukfisz, B.; Frydrysiak, M.; Prochoń, M. Integration of Safety Aspects in Modeling of Superheated Steam Flash Drying of Tobacco. *Energies*, **2021**, *14*(18). https://doi.org/10.3390/en14185927.

[225] Hamawand, I.; Yusaf, T.; Bennett, J. Study and Modelling Drying of Banana Slices under Superheated Steam. *Asia-Pacific J. Chem. Eng.*, **2014**, *9*(4). https://doi.org/10.1002/apj.1788.

[226] Fu, H.; Zheng, Z.; Ren, L.; Yang, P.; Xu, J.; Xie, W.; Yang, D. Simulation and Experimental Study on the Effect of Deformation on Heat and Mass Transfer in Potato Chips during Superheated Steam Drying. *J. Food Process Eng.*, **2024**, *47*(1). https://doi.org/10.1111/jfpe.14518.

[227] Le, K. H.; Tran, T. T. H.; Kharaghani, A.; Tsotsas, E. Modeling of Superheated Steam Drying of Wood Particles. *J. Mech. Eng. Res. Dev.*, **2020**, *43*(1), 160. https://doi.org/10.1002/ceat.202000133.

[228] Betoret, E.; Calabuig-Jiménez, L.; Barrera, C.; Rosa, M. D. Sustainable Drying Technologies for the Development of Functional Foods and Preservation of Bioactive Compounds. In *Sustainable Drying Technologies*; del Real, O. J., Ed.; Intech Open, **2016**; pp. 38–56. https://doi.org/10.5772/64191.

[229] Sobulska, M.; Wawrzyniak, P.; Woo, M. W. Superheated Steam Spray Drying as an Energy-Saving Drying Technique: A Review. *Energies*, **2022**, *15*(22), 8546. https://doi.org/10.3390/EN15228546.

[230] Superheated Steam Drying for Sustainable Recyclable Web-Like Materials Is Expected to Reduce, on the Entire Production Line. https://cris.vtt.fi/en/projects/superheated-steam-drying-for-sustainable-recyclable-web-like-mate#:~:text=SSD (accessed April **18**, 2024).

[231] *Industrial Dryers: Global Strategic Business Report*; Dublin, Ireland, **2024**.

[232] *Global Industrial Dryers Market by Product (Conveyor Dryers, Drum Dryers, Flash Dryers), End-User Industry (Cement, Chemicals, Fertilizer)-Forecast 2024-2030*; Dublin, Ireland, **2024**.

[233] *Industrial Dryers Market by Product Type (Direct, Indirect, and Others), Type (Rotary, Fluidized Bed, Spray, and Others), Application (Food, Pharmaceutical, Fertilizer, Chemical, Cement, and Others), and Region 2024-2032*; New York, United States, 2023.

[234] Martynenko, A.; Alves Vieira, G. N. Sustainability of Drying Technologies: System Analysis. *Sustain. Food Technol.*, **2023**, *1*(5), 629–640. https://doi.org/10.1039/d3fb00080j.

[235] Kumar, K. R.; Dashora, K.; Kumar, S.; Dharmaraja, S.; Sanyal, S.; Aditya, K.; Kumar, R. A Review of Drying Technology in Tea Sector of Industrial, Non-conventional and Renewable Energy Based Drying Systems. *Appl. Thermal Eng.*, **2023**, *224*, 120118.

[236] Christian, M.; Christian, B. High Temperature Convective Drying of Softwood and Hardwood: Drying Kinetics and Product Quality Interactions. In: *Proc. Fourth Int. Drying Symp.;* Toei, R., Mujumdar, A. S., Eds.; Session I, Paper 13, Japan Soc. Chem. Engrs.: Kyoto, Japan, **1984**; pp. 662–666.

# Index

Printed in the United States
by Baker & Taylor Publisher Services